Cristiane Carine Santos

Concrete with rice husk ash and foundry sand

Cristiane Carine Santos

Concrete with rice husk ash and foundry sand

Concrete study

Imprint

Any brand names and product names mentioned in this book are subject to trademark, brand or patent protection and are trademarks or registered trademarks of their respective holders. The use of brand names, product names, common names, trade names, product descriptions etc. even without a particular marking in this work is in no way to be construed to mean that such names may be regarded as unrestricted in respect of trademark and brand protection legislation and could thus be used by anyone.

Cover image: www.ingimage.com

This book is a translation from the original published under ISBN 978-613-9-62414-0.

Publisher:
Sciencia Scripts
is a trademark of
Dodo Books Indian Ocean Ltd. and OmniScriptum S.R.L publishing group

120 High Road, East Finchley, London, N2 9ED, United Kingdom
Str. Armeneasca 28/1, office 1, Chisinau MD-2012, Republic of Moldova, Europe
Printed at: see last page
ISBN: 978-620-7-72632-5

SUMMARY

I dedicate this achievement to my parents, Neri and Marilei, my brother Marciano, and my boyfriend Gustavo, for their support, encouragement and understanding throughout all these years.

ACKNOWLEDGMENTS

First of all, I would like to thank God for always having enlightened my steps, for leading me in the right direction and for giving me the strength and courage to pursue my goals.

To my parents, Neri José dos Santos and Marilei Elizabeth dos Santos, and my brother Marciano José dos Santos, for their encouragement, strength, understanding and support, who have been by my side all these years helping me to achieve this feat, which is also yours, and to say that without you I wouldn't have got as far as I did. And on their behalf, I would like to thank the rest of my family who have always supported me.

To my boyfriend Gustavo Isac de Freitas, for his love, patience, understanding and support, always willing to help me. Even though he didn't have much knowledge of the field, he never failed to pay attention to my university studies, on the contrary, he always made an effort to contribute.

To my advisor, Professor Diorges Carlos Lopes, for his attention, teachings and dedication in carrying out this work.

To Professor Cristiana Eliza Pozzobon for her professional example, dedication and shared knowledge.

To Professor Lucas Fernando Krug, for making himself available to discuss the research, analyze the results, for his attention and dedication.

To Professor Bóris Sokolovicz, for his efforts in trying to schedule the Scanning Electron Microscopy test.

To the other teachers and friends who, in one way or another, contributed to this achievement.

To the LEC laboratorians, Luis Donato, Tulio Ribeiro and Roberto Brandao, for their help during the tests, for their attention and dedication.

To my research colleagues, Pedro Goecks, Gabriela Blatt, Felipe Dacanal dos Anjos and Tatiane Thomas Soares, I thank you for your commitment, dedication, friendship and all your help. And on their behalf, I would like to thank all my colleagues in the PET Group who were always there to lend a helping hand when I needed it.

To all the colleagues and friends I made during these five years at university, for the days of study, work and projects. Especially Tagiana Krenchinski and Paulo Dobler da Costa, for the friendship that came to fruition, for their patience on the eve of exams and when assignments were due. They are people I will carry in my heart forever.

FUNDIMISA, for donating the foundry sand.

To FAURO MATERIAIS DE CONSTRUÇÂO for their help in unloading the foundry sand.

To all those who have not been mentioned, but who have somehow been by my side and helped me to complete this work.

Thank you so much!!!

Without dreams, life is dull. Without goals, dreams have no foundation. Without priorities, dreams don't become real. Dream, set goals, establish priorities and take risks to make your dreams come true. It's better to err on the side of trying than to err on the side of omitting!

Augusto Cury

SUMMARY

SANTOS, C. C. **Strength Analysis of Concrete with Partial Substitution of Cement by Rice Husk Ash Microsilica and Partial Substitution of Natural Sand by Foundry Sand.** 2015. Course Conclusion Paper. Civil Engineering Course, Universidade Regional do Noroeste do Estado do Rio Grande do Sul - UNIJUi, Ijui, 2015.

Care and concern for the environment is becoming increasingly intense, and it is necessary to find alternatives that help minimize the impacts that are continually being caused. Through research and studies, alternatives have emerged for reusing waste that can increase the strength of concrete and contribute to sustainability, instead of being disposed of inappropriately. As rice production is intense in Rio Grande do Sul and other states, a large volume of solid waste is generated as a result of processing and beneficiation activities, identified as rice husk ash. When this waste is not disposed of correctly, it causes serious environmental and health problems. This has led to the search for studies and research into the possibility of substituting these residues as building materials, which in addition to generating ecological and economic benefits, can also improve some of the properties of concrete. As well as rice husk ash, there are numerous wastes that can be reused to replace the materials that make up concrete, including foundry sand, which is a waste generated by foundry industries. Sand serves as a mold for metal parts, is used in all casting processes and is responsible for shaping the metal. After going through various processes, this sand must be discarded and replaced with new sand. As it is no longer possible to reuse it, it must be sent to an industrial landfill, which requires care and monitoring, otherwise it creates serious problems for the industry responsible and the environment. Some time ago this was the only alternative destination for industrial sands, but through studies and research new ways of reusing this waste have emerged, such as partial substitution for materials that make up concrete. Taking these aspects into account, this study sought to partially replace cement with rice husk ash and natural sand with foundry sand in concrete, in order to reduce environmental impacts, production costs and, above all, improve the properties of the concrete. From the substitutions of these residues, which were carried out by dosing the concrete using the ABCP Method, it was possible to analyze the influence provided to the concrete by the substitutions of 5%, 10%, 15% and 20% of each of the residues in separate mixtures and in a single mixture with the two residues. The compressive strength at 7, 14, 21 and 28 days and the tensile strength at 28 days of age of the concrete were also evaluated. The results showed that the mixes with only foundry sand had an increase in strength, and the mixes with only rice husk ash and with the two residues together remained close to the reference mix, but slightly lower. Even so, it can be said that it is feasible to reuse these residues in the construction industry, as in addition to improving the properties of concrete, they also contribute to reducing the consumption of raw materials, helping to preserve the environment.

Keywords: Solid waste. Mineral additions. Residual sands. Environmental problems.

1 INTRODUCTION

Increasingly, care for the environment must be intensified, and one of the alternatives that has made a great contribution is the use of waste that can increase the properties of concrete and consequently help with sustainability, instead of disposing of it inappropriately (KELM, 2011).

According to Metha (1994), the estimated worldwide consumption of concrete is 5.5 billion tons per year, and consequently the concrete industry is the largest consumer of natural reserves, thus requiring the incorporation of technologies to reduce the environmental problems it generates.

In the state of Rio Grande do Sul, the economy is strongly linked to agricultural activities. Among these is the production of a large volume of rice, which ends up generating a huge amount of solid waste derived from processing and beneficiation activities, identified as rice husks and the ash resulting from these husks. However, this waste is classified as a major source of pollution and contamination, and when not properly managed and disposed of it causes serious impacts on the environment and people's health (IRGA, 2013).

Due to the consequences generated by the large number of solid residues, several studies have been carried out on their reuse in concrete, and the feasibility of partially replacing rice husk ash (RHA) in concrete has already been proven, with satisfactory results such as: increased compressive strength, greater concrete durability and cement savings. In this context, the use of CCA becomes one of the solutions to the environmental problem caused by the disposal of this waste, reducing the areas needed for agro-industrial disposal and minimizing the risks of environmental pollution (POUEY, 2006).

Other solid waste that also causes concern is related to the foundry industry. This waste is mainly made up of so-called residual sands or foundry sands.

When the sands from the foundry process become unusable, the waste must be disposed of and deposited in landfills, following all the rules established by the State Environmental Protection Foundation (FEPAM). These landfills require a lot of care and monitoring, otherwise they generate numerous environmental problems, due to the large volume of waste and especially in relation to the cost of companies maintaining and building new trenches. As a solution to the large volume of these sands, it is necessary to manage them properly with a view to alternatives for reuse (LIMA, 2014).

However, among the various studies and also with technological advances, we realize that the construction industry has the capacity to provide a solution to the waste generated in this field, both rice husk ash and foundry sands, with the possibility of including them in the construction industry as building materials. This would make it possible to reduce the cost of products and, above all, reduce the amount of waste emitted into the environment.

Taking these aspects into account, this research was based on the study of the possibility of reusing rice husk ash microsilica and foundry sand in Portland cement concrete.

The aim of this study was to analyze the properties of concrete by partially replacing rice husk ash with microsilica and partially replacing it with foundry sand in the proportions of 5%, 10%, 15% and 20%, first in separate mixtures and then with the two residues in the same concrete mix. To try to provide a technically and economically viable solution through the reuse of these residues.

Concrete dosage tests were carried out using the ABCP dosage method. The Compressive Strength and Diametral Compressive Tensile Strength properties of the different replacements of cement with rice husk ash microsilica and natural sand with foundry sand were analyzed.

2 LITERATURE REVIEW

This stage will present concepts obtained through intensive studies of work that is analogous to the objective of the proposed work.

2.1 CONCRETE

According to the magazine Concreto & Construçoes (2009), published by the Brazilian Concrete Institute - IBRACON, concrete is a widely used construction material. It is found in our masonry houses, on highways, in bridges, in buildings, in towers, in hydroelectric and nuclear power plants, in sanitation works, among other places.

According to Yazigi (2009), the main binder used in concrete mixing is cement, which, when in contact with water, produces exothermic crystallization of hydrated products, acquiring mechanical strength.

Salgado (2009) mentions that the largest constituent materials of concrete are aggregates, classified as fine and coarse aggregate, and it is therefore important to choose quality aggregates to generate good performance in the production of concrete, both in the production of mix by means of a batching plant or by hand.

The estimated consumption per year is 11 billion tons of concrete, which, according to the Ibero-American Federation of Precast Concrete (FIHP), gives an average consumption of 1.9 tons of concrete per inhabitant per year, less than the consumption of water alone.

According to Inês Battagin, superintendent of CB-18 of the Brazilian Association of Technical Standards (ABNT), described in the magazine Concreto & Construçao (2009), concrete is a homogeneous mixture of cement, small and large aggregates, with or without the incorporation of minor components (chemical additives and additions), which develops its properties by hardening the cement paste. The author also mentions that concrete is a material that makes it possible to build structures in a wide variety of shapes and different types of environments, because as well as being durable, it also has the ability to withstand stress.

According to Mehta and Monteiro (1994), concrete is classified according to its specific mass:

- Lightweight concrete (specific mass less than 1800 Kg/m);[3]

- Normal concrete or conventional concrete (specific mass between 2300 and 2400 Kg/m3),

- Heavy concrete (specific mass greater than 3200 Kg/m3),

According to Isaia (2005), countless techniques and compositions are known and studied so that their applications can be adjusted in technical, economic and social terms, paying attention to the use of finite resources and the reduction of pollution. Over time, and with advances in technology and materials, the use of different materials that can form part of the composition of concrete has become more widespread, leading researchers to study alternatives for combining durability with sustainability.

2.1.1 Considerations on Portland cement

According to the Brazilian Portland Cement Association (ABCP), cement can be described as a fine powder with agglomerating, binding or binding properties, which hardens under the action of water. When in the form of concrete, it can take on different shapes and volumes, depending on its usefulness in various types of construction work.

According to Mehta (1999), one of the main components among the materials that make up concrete is Portland cement. Due to the fact that it is used in considerable quantities everywhere, this has caused serious concern for the population, because during the cement production process, elements are emitted that cause undesirable changes to the environment. Cement production contributes approximately 5% of the amount of CO_2 (carbon dioxide) emitted into the atmosphere each year.

According to ABCP, looking back to the 1990s, in 2008 Brazil had the lowest rate of CO_2 emissions per ton of cement (560 kg of CO2 per ton of cement).

According to Mehta and Monteiro (2008) Portland cement concrete is the most consumed manufactured material in the world. Taking this and other aspects into account, it is of great importance to study alternative materials that can partially replace cement in a concrete mix, with the aim of preserving the environment and making it more economically viable.

2.1.2 Constituent materials

- Portland cement

Portland cement is the name given to the world-famous and extremely important material in the construction industry.

Standard NBR 11578/1997 defines composite Portland cement as:

> Hydraulic binder obtained by grinding Portland clinker to which the necessary quantity of one or more forms of calcium sulphate is added during the operation. During grinding, it is permissible to add pozzolanic materials, granulated blast furnace slag and/or carbonate materials to this mixture.

According to ABCP's Basic Guide to the Use of Portland Cement (2002), Portland cement was created by an English builder, Joseph Aspdin, who patented it in 1824. At the time, it was common in England to build with stone from Portland, an island in the south of England. As the result of Aspdin's invention was similar in color and hardness to this Portland stone, he wrote that name on his patent, which became known as Portland cement.

According to ABCP's Basic Guide to the Use of Portland Cement (2002), the manufacture of cement depends on the main materials: limestone, clay, iron ore and gypsum. During the manufacturing process, the materials are analyzed several times in order to achieve the desired chemical composition. Table 1 shows the chemical composition of ordinary and composite Portland cements.

Table 1 - Composition of ordinary and composite Portland cements

Type of portland cement	Acronym	Composition (% by mass)				Brazilian Standard
		Clinker + gypsum	Granulated blast furnace slag (acronym E)	Pozzolanic material (acronym Z)	Carbon material (acronym F)	
Common	CP I	100	-			NBR 5732
	CP I-S	99-95	1-5			
Compound	CP II-E	94-56	6-34	-	0-10	NBR 11578
	CP II-Z	94-76	-	6-14	0-10	
	CP II-F	94-90	-	-	6-10	

Source: Basic guide to the use of Portland cement, 2002.

Tables 2 and 3 show the compositions of blast furnace and pozzolanic portland cements and the compositions of high initial strength portland cement, respectively.

Table 2 - Composition of blast furnace and pozzolanic Portland cements

Type of portland cement	Acronym	Composition (% by mass)				Brazilian standard
		Clinker + gypsum	Granulated blast furnace slag	Pozzolanic material	Carbon material	
Blast furnace	CP III	65-25	35-70	-	0-5	NBR 5735
Pozzolanic	CP IV	85-45	-	15-50	0-5	NBR 5736

Source: Basic guide to the use of Portland cement, 2002.

Table 3 - Compositions of high initial strength Portland cement

| Type of portland cement | Acronym | Composition (% by mass) | | Brazilian Standard |
| | | Clinker + gypsum | Carbon material | |

| High initial resistance | CP V-ARI | 100-95 | 0-5 | NBR 5733 |

Source: Basic guide to the use of Portland cement, 2002.

Revista Concreto & Construçoes (2009) classifies the different types of cement Portland as:

Ordinary Portland Cement (CP I): It is used to control setting (time required for the partial hardening of the compound). It is recommended for concrete construction in general, when no special cement properties are required.

Portland Composite Cement (CP II): Fully satisfies the needs of most common applications and in many cases has additional advantages.

High Kiln Portland Cement (CP III): It has greater impermeability and durability, low heat of hydration and high resistance to expansion and sulfates (alkali-aggregate reactions), this cement is useful in concrete-mass works, such as the construction of dams.

Pozzolanic Portland cement (CP IV): Suitable for works exposed to the action of running water and aggressive environments due to its properties of low permeability, high durability, high compressive strength at advanced ages.

High Initial Strength Portland Cement (CP V - ARI): Achieves high strengths in the first few days of application, it is used in the manufacture of masonry blocks, paving blocks, pipes, slabs, curbs, posts and pre-cast architectural elements, which require a high initial strength cement for rapid deformation.

- Mineral aggregates

Bauer (2000) defines aggregate as a particulate material with practically zero chemical activity, made up of mixtures of particles covering a wide range of sizes. Since small and large aggregates are found in nature in considerable quantities and have a low unit value, their consumption is an important indicator of a country's socio-economic profile.

According to the Manual of Aggregates for Construction (2009), the mineral substances most consumed in the construction industry are aggregates, and also the most significant in terms of the quantity consumed in the world.

According to the Fundaçao Instituto de Pesquisas Econômicas da Universidade de Sao Paulo (FIPE), available on the website of the Associaçao Nacional das Entidades de Produtores de Agregados para Construçao Civil (ANEPAC), the average consumption of aggregates in Brazil is shown in Table 1:

Table 1 - Number of aggregates in Brazil

Self-build units up to 35 m²	*21 t of aggregates*
Affordable housing of 50 m²	*68 tons of aggregates;*
Maintenance of municipal roads	*100 t/km*
Roads	*3 thousand t/km*
Urban paving	*varies from 0.116 m³/m² to 0.326 m³/m²*
1,200 m scale²	*1000 m³ of aggregates (1,680 t)*

Source: Adapted from ANEPAC. Inventta elaboration.

According to Collins and Fox (1985), the classification of aggregates must take into account the following information: the origin of the material (natural or artificial aggregates), the petrographic class or name, as well as the age of the rock, color, granulometry and fissility.

According to Helene (1992) and Bauer (2000), aggregates are classified according to:

a)   The origin:

>   Natural: Aggregates that are in particulate form in nature, such as sand and gravel;

>   Industrialized: Aggregates where their composition has been achieved by industrial processes, in this context, the raw material can be: rock, blast furnace slag.

b)   The size of particles:

>   Fine aggregate: sand;

>   Coarse aggregate: gravel and crushed stone.

Standard NBR 7211 (2009) describes the characteristics of aggregates, small and large, of natural origin, found in fragments or resulting from the crushing of rocks, as:

> Sand or fine aggregate is sand of natural origin or resulting from the crushing of stable rocks, or a mixture of both, whose grains pass through the 4.8 mm ABNT sieve and are retained on the 0.075 mm ABNT sieve. And it defines coarse aggregate as pebbles or gravel from stable rocks, or a mixture of both, whose grains pass through a square mesh sieve with a nominal opening of 152 mm and are retained on the 4.8 mm ABNT sieve.

2.1.3 Concrete strength

Mehta & Monteiro (1994) define strength as the factor responsible for a material's ability to withstand tension without rupturing. In concrete, strength is related to the stress required to cause fracture and is similar to the degree of rupture at which the applied stress reaches its maximum value.

According to Helene & Terzian (1992), compressive strength is the characteristic of concrete that best qualifies it, but for this it is necessary to take into account dosage and preparation, as well as durability and workability properties.

According to Kelm (2011), the strength of concrete is obtained from the hydration of cement, which is a very slow process. Most of the time, concrete strength tests are carried out by molding specimens, which are cured under specific conditions of temperature and humidity, and broken to obtain the compressive strength result after 28 days of age.

Mehta and Monteiro (1994) classify concrete by its compressive strength as:

- Low-strength concrete (compressive strength less than 20 MPa);

- Moderate strength concrete (compressive strength between 20 and 40 MPa);

- High-strength concrete (strength greater than 40 MPa).

For Neville (1997), even if the concrete is not designed to resist tension, carrying out this test is important because it can estimate the load at which cracking will occur. The absence of cracks is necessary for the preservation of a concrete structure, and also to prevent corrosion of the reinforcement.

2.2 MINERAL ADDITIONS

According to Mehta and Monteiro (1994), mineral admixtures are finely ground siliceous materials added to concrete in considerable quantities, most often in the range of 20 to 100% of the Portland cement mass.

Metha and Monteiro (2008) mention that mineral additions have favorable effects on concrete, mainly by reducing the porosity of pastes and refining the grains of calcium hydroxide. The change in microstructure generated by the reduction in porosity leads to an increase in compactness, making it more difficult for aggressive agents to access the interior of the concrete. Metha and Monteiro (2008) also present the types of mineral additions and their classifications in Table 2.

Table 2 - Types and classifications of mineral additions

Classification	Types of Additions
Cementing	Granulated blast furnace slag
Cementing and Pozzolanic	Fly ash with a high calcium content

	Active silica
Superpozolanas	Metakaolin
	Rice husk ash
Pozzolans with some	Low calcium fly ash Calcined clays
	Natural materials (volcanic and sedimentary or| gem)
Poorly reactive pozzolans	Slowly cooled blast furnace slag
	Kiln ashes
	Boiler slag
	Burnt rice straw in the field
Inert fillers	Calcium dust, stone dust

Source: Mehta and Monteiro, 2008

Isaia (1995) states that pozzolanicity and fineness are the main properties that affect the performance of mineral additions, as well as the quantity of pozzolan and its granulometric composition.

According to Sokolovicz (2013), pozzolans are classified as natural (volcanic origin) and artificial, which are derived from the calcination of rocks, coal-burning residues and agro-industrial residues such as fly ash and rice husk ash.

Numerous studies have shown that it is possible to use different mineral additions, in various levels and combinations, to create new concrete mixtures.

According to Neville (1997), the use of mineral admixtures in concrete is justified by their technical advantages, as well as their energy and environmental advantages. Among these additions are pozzolans, such as rice husk ash, which is an agro-industrial residue produced by burning rice husks, which is used as a source of energy in the processing of the grain.

According to Frizzo (2001), mineral admixtures can be introduced into concrete by partially replacing cement, by adding to it, increasing the amount of binder material, or as a fine aggregate. However, in order for mineral admixtures to be introduced into concrete, they must meet certain requirements, such as: fineness, appropriate particle size, and their physical and chemical characteristics must be known.

2.2.1 Rice husk (CA)

Rice husks are a residue removed during the rice grain refining process and originate from agricultural waste. One of the major problems for rice producers is storing the residue generated, due to its low commercial value or interest in using it in agriculture (GONÇALVEZ, 2009).

According to Houston (1972), approximately 20% of the grain's weight is CA, making it the largest by-product of grain production. Hulls are made up of 50% cellulose, 30% lignin and 20% organic residues. Figure 1 illustrates rice husks.

Figure 1 - Rice husk

Source: Universoagro (2013)

2.2.2 Rice husk ash (RHA)

According to Milani (2008), CCA is generated by burning rice husks in furnaces, open fires and ovens that allow temperature control.

According to Mehta & Monteiro (1994), each ton of paddy rice produces around 200 kg of husk, which by combustion form 40 kg of ash. As considerable quantities of this waste are generated in various parts of the country, studies and research into its reuse are extremely important, both in the construction industry and in other areas.

Several studies on the use of rice husk ash as a partial substitute in concrete have shown that it is feasible and brings positive results. According to Kelm (2011), the use of CCA is one of the solutions to the disposal of this material, which has been causing serious environmental problems, as it reduces the areas needed to dispose of this agro-industrial waste and also minimizes the risks of environmental pollution.

According to Mehta (1992) apud Missau (2004), rice husk ash is classified as a highly reactive pozzolan, consisting mainly of silica in non-crystalline form.

The term pozzolan, according to NBR 12653 (2015), is defined as a material that has little or no agglomerating activity, but when finely divided and in the presence of water, reacts with calcium hydroxide at room temperature to form compounds with agglomerating properties.

According to Guedert (1989), the silica contained in the sample is influenced by the temperature at which the CA is fired. For maximum reactivity, the temperature must be controlled, with temperatures between 300 and 800 degrees Celsius.

According to Tashima (2006), no matter how rice husks are burned, the resulting ash has a silica content of around 74% to 97%. The type of burning mainly influences the morphology of the silica present in the ash. Whether the silica appears in an amorphous state (more reactive) or in a crystalline state depends on the temperature reached during burning.

According to Santos (1997), taking into account that the volume of rice husk ash is produced in large quantities, companies do not have suitable places to dispose of this waste, and end up throwing it in inappropriate places, as shown in Figure 2, where it is deposited next to a highway.

Figure 2 - Improperly disposed of rice husk ash

Source: SANTOS, S. (1997)

2.2 studies WITH RICE HUSK ASH MICROSILICA in concrete

2.3.1 Influence of CCA on concrete strength

The studies on rice husk ash microsilica carried out by Kelm (2011) show that, with the use of two a/ag factors, at the age of 7 days the compressive strength of all the mixes is lower than that of the reference mix, cast without additions, as shown in Figure 3.

Figure 3 - Compressive strength at 7 days

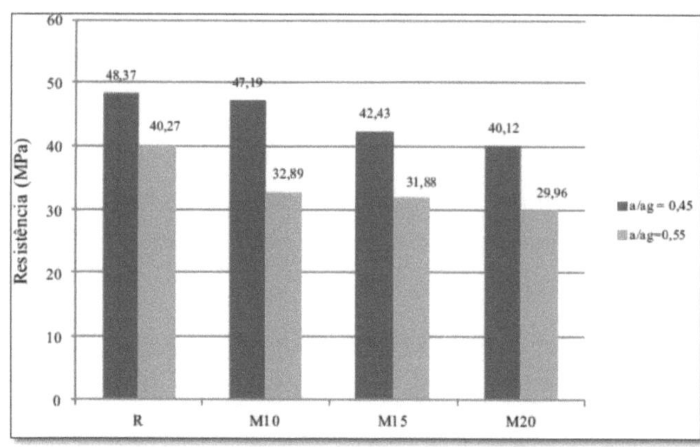

Source: Kelm, 2011

Also according to Kelm (2011), at 28 days of age, the compressive strength of the concrete increased compared to 7 days of age, and all the mixes with an a/ag factor equal to 0.45 exceeded the strength values of the reference mix, as shown in Figure 4.

Figure 4 - Compressive strength at 28 days

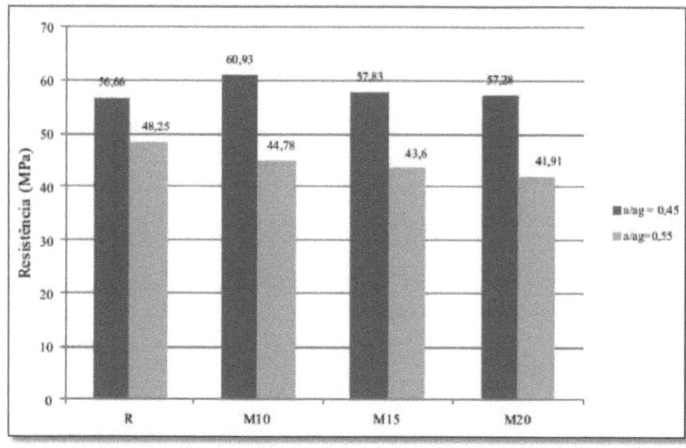

Source: Kelm, 2011

Ludwing (2014) showed in his research, with cement replaced by rice husk ash microsilica, that at the age of 7 days, the concrete with 3% and 7% replacement had higher strengths than the reference mix, and only the mix with 5% replacement had lower strength, as shown in Figure 5.

Figure 5 - Concrete strength at 7 days

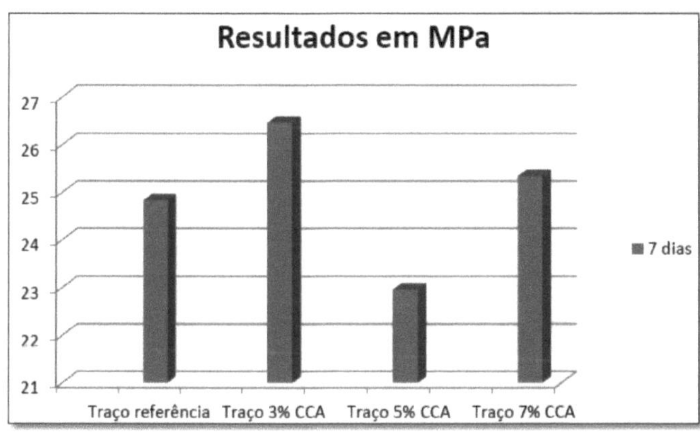

Source: Ludwing, 2014

After 28 days of age, according to Ludwing (2014), the strength of concrete with a 3% substitution of rice husk ash microsilica was higher than that of concrete with no substitution of residues, while the 5% and 7% substitutions had lower strengths, as can be seen in Figure 6.

Figure 6 - Concrete strength at 28 days

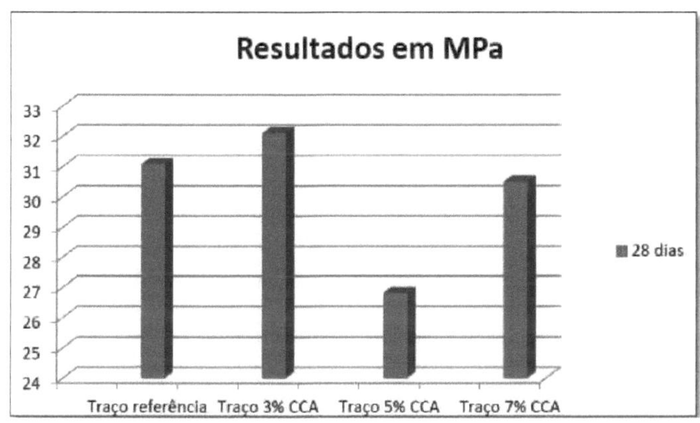

Source: Ludwing, 2014

Ludwing (2014) concluded from the compressive strength results of replacing 3% of the cement with rice husk ash microsilica that this mixture can be recognized by the construction market, since the strength exceeds that of concrete cast without the addition of waste.

2.4 FOUNDRY SAND

One of the by-products of the foundry industry is foundry sand made from ferrous and non-ferrous materials. Silica sand is used to make molds, cores and nubs for metal parts, and is recycled and reused many times during the process. When the sand is at an unserviceable stage for the foundry process, it is discarded and referred to as "foundry sand" (SIDDIQUE; SINGH, 2011).

According to Scheunemann (2005), the main waste generated during the demoulding of metal parts in the foundry process is foundry sand. It is thanks to more versatile manufacturing methods and the shapes and sizes of metal parts that the foundry industry is gaining more attention than other industries.

The physical and chemical characteristics of foundry sand depend on the type of foundry process to which they are previously subjected (SIDDIQUE; SINGH, 2011).

According to NBR 10004 (2004), foundry sand, which is used to make molds for metal parts, is classified as non-hazardous waste, but depending on each situation or process, it can have some pollutants, such as metals (iron, aluminum, nickel, chrome, lead, zinc, etc.) and phenolic resins, which can affect the environment when disposed of or handled improperly.

Matos and Schalch (1997) define that the first step in the molding process is to prepare the molding sand, which is made up of base sand, residual sand, recycled materials and additives, which are placed in a mixer. Using this molding sand, the molds are prepared, at which point the cores are also placed, which are made up of the same materials as the molds, and are responsible for generating holes and recesses in the metal parts that will be produced. Then, after the molds and cores have been produced, the metal is melted and the molds are filled by pouring the molten metal. In the final stage, the parts are demoulded and finished.

According to Peixoto (2003), after being reused numerous times in the casting process, the sand starts to accumulate contaminating materials, such as resins, coals and other substances, which can damage the quality of the molds. To prevent this from happening, the residual sand must be replaced with new sand during the molding process. These sands that can no longer be used are discarded and become foundry sand.

One of the major problems faced by foundry companies, cited by Lopes (2009), is related to the disposal of foundry sands, because when they are sent to landfills, they end up mixing with other waste that is generally more contaminating, creating risks for foundry companies, which are held responsible for the environmental damage caused.

2.4.1 Casting process

Gossen (2005) describes all the casting processes used to make the parts:

- Molding: this is the beginning of the process, in which a mold of the part to be produced is made and used in the production of the other molds. When using sand bound with clay (green sand), the sand is compressed around the mold until it reaches the desired hardness. And when chemical ingredients are used, the mold is chemically hardened after light mechanical compaction.

-       Melting: this is where the production of the casting begins. This process generates waste such as furnace slag, dust from the furnace exhaust system, scrap metal unsuitable for the process and raw material dust.

-       Machining: this is where the sand cores, known as "cores", are made. These cores are made of ingot sand, which has sufficient strength and hardness to be fixed inside the mold.

-       Demoulding: the demoulding process begins after the part has cooled. It is a stage in which the mold must be broken to remove the molten component. This generates waste such as iron splashes, molding sand, sand fines and core residue.

-       Channel breakage: the casting feed system breaks down. The waste generated is scrap and molding sand.

-       Finishing: process in which the grooves are cut and finished using rotating wheels and other tools to improve the visual appearance.

-       Heat treatment: this stage is used to bring the structure of the material into line with its specifications. This treatment aims to improve the mechanical and corrosive resistance of the materials.

-       Oiling/Painting: the aim is to protect the casting by painting or oiling.

Figures 7 and 8 show the sands before and after the casting processes.

Figure 7 - Natural sand before going through the casting process

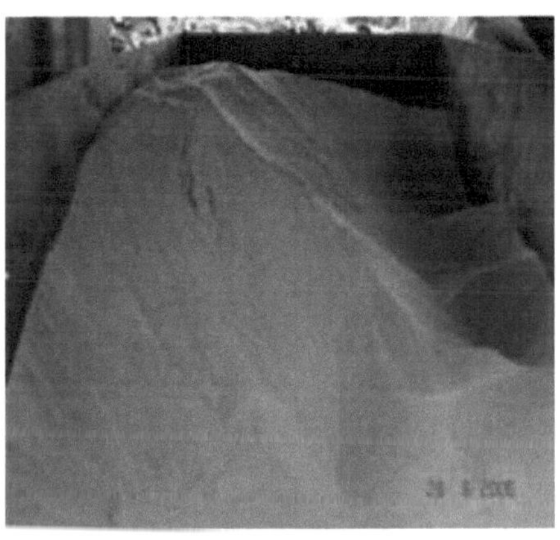

Source: Albrecht website (2013)

Figure 8 - Sand after the casting process

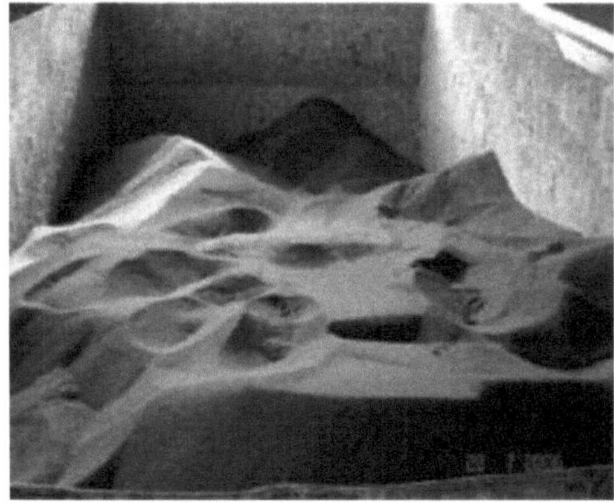

Source: Albrecht website (2013)

2.5 STUDIES ON FOUNDATION SAND IN CONCRETE

2.5.1 Influence of foundation sand on concrete strength

Studies carried out by Lima (2014), with the substitution of natural sand by foundry sand, show that all the mixes molded with the substitution of waste have higher strengths than the reference mix, at 7 days of age of the concrete, as shown in Figure 9.

Figura 9   - Compressive strength at 7 days

Source: Lima, 2014

With a longer curing time, Lima (2014) showed even higher compressive strength values at 28 days than at 7 days, as shown in Figure 10.

Figura 10   - Compressive strength at 28 days

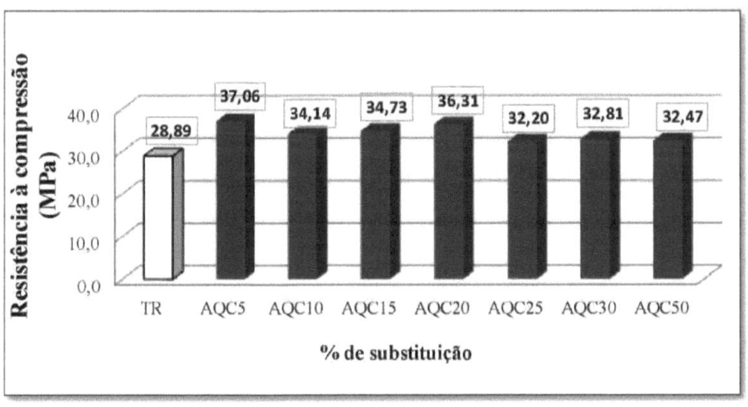

Source: Lima, 2014

Lima (2014) emphasizes in his research that foundry sand has finer grains than natural sand, which is why it fills the voids in concrete better, contributing to better hydration and increased strength.

2.6 SOLID WASTE

According to NBR 10004 (2004), solid and semi-solid waste are those that derive from the activities of the community of origin: industrial, domestic, hospital, commercial, agricultural, services and sweeping.

Although the foundry industry uses scrap metal as a raw material, it produces large quantities of solid waste, including slag, molding sand and various kinds of dust (DANTAS, 2003).

According to data from the Brazilian Foundry Association (ABIFA), foundry sands are currently one of the industrial solid wastes with the highest production volume in Brazil. When this waste is not disposed of correctly, it can cause serious environmental problems.

According to NBR 10004 (2004), waste is classified as:

a)   Class I - Hazardous waste: This can present risks to public health and the environment due to its physico-chemical and infectious properties. Hazardous waste is considered to have at least one of the following characteristics: flammability, corrosivity, reactivity, toxicity and pathogenicity. Waste classified as hazardous requires special disposal precautions.

b)   Class II - Non-hazardous waste: does not have any of the above characteristics, but can be classified into two subtypes:

Class II A - Non-Inert: those that do not fit into the previous item, Class I, or the next item, Class II B. They generally have one of the following characteristics: biodegradability, combustibility and solubility in water.

23

Class II B - Inert: any waste that, subjected to static or dynamic contact with water, does not have any of its components solubilized at concentrations above the standards for water portability, with the exception of color, turbidity, hardness and taste.

According to Lima (2014), through leaching tests, which are procedures used to analyze the potential for transferring materials to the natural environment, foundry sand can be classified as a **Class II non-hazardous waste, with** no chemical element exceeding the values allowed by NBR 10004 (2004).

FEPAM (2003) presented data on the generation and disposal of hazardous waste (Class I) and non-inert or non-hazardous waste (Class II) from industries located in the state of Rio Grande do Sul. The data was collected in 2002 and shows the potential for generating solid industrial waste. Table 3 and figure 11 show the amount of solid industrial waste generated by waste class, i.e. Class I or hazardous waste and Class II or non-hazardous (non-inert) waste.

Table 3 - Distribution of the amount of industrial solid waste generated by Class

WASTE CLASS	QUANTITY INFORMED VIA THEINVENTORY	QUANTITY REPORTED via the plans	TOTAL IN VANE
Class 1	182 170	7.033	189.203
Class 11	⅛16 300	1 227.782	2.174.682
Total	1.129.070	1.234.816	2.363.886

Source: Fepam, 2003

Figure 11 - Industrial Solid Waste Generation by Class

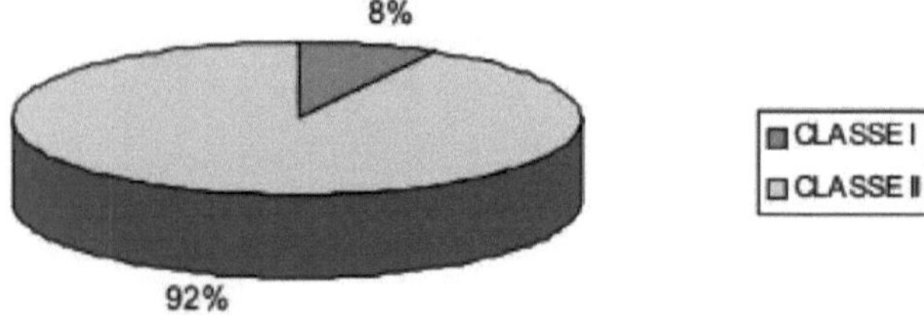

Source: Fepam, 2003

Figures 12 and 13 show the amount of Class I hazardous industrial solid waste and the amount of Class II non-hazardous industrial solid waste, respectively, generated by industrial sector.

Figure 12 - Generation of Class I Industrial Solid Waste by industrial sector

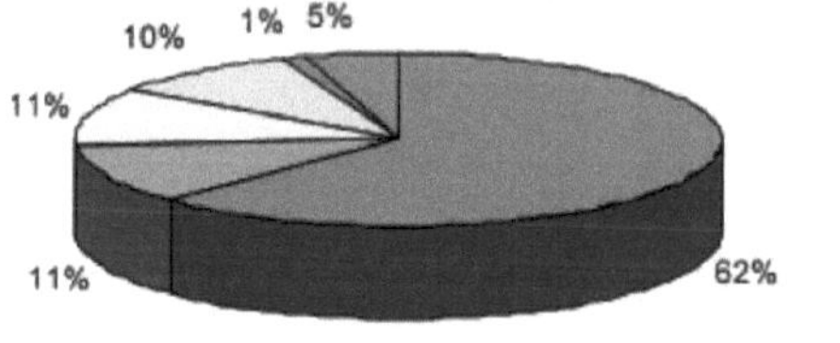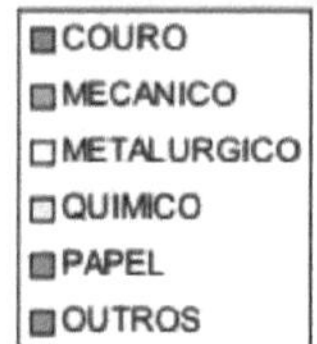

Source: Fepam, 2003

Figure 13 - Generation of Class II Industrial Solid Waste by industrial sector

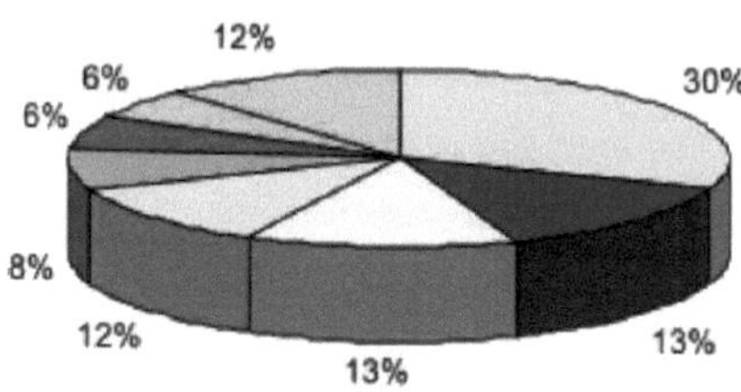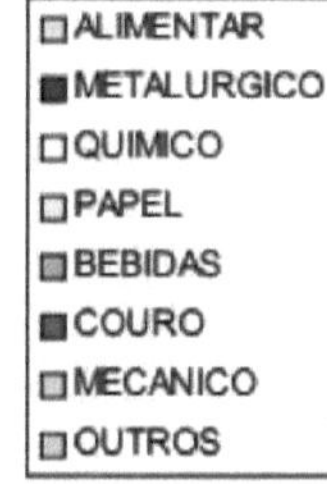

Source: Fepam, 2003

According to data from ABIFA (2015), one of the largest quantities of solid industrial waste produced in Brazil is foundry sand. This has led to important studies regarding the reuse of this waste in concrete mixtures, mainly due to the enormous volume generated and the consequences for human beings.

3 METHODOLOGY

This chapter presents the steps taken from obtaining the materials for the research to the tests, standards and characterization results.

3.1 SEARCH STRATEGY

This research can be classified as applied, and in terms of procedures, it is a quantitative research, developed through laboratory tests.

3.2 RESEARCH PLANNING

Firstly, research and studies were carried out to gain a better understanding of the subject in question, developing the theoretical foundation. The second stage was based on the acquisition of materials, such as: medium sand, foundation sand, cement, rice husk ash and gravel 1. After preparing these materials, their characterization was carried out, one of the main procedures for obtaining the concrete dosage mix.

Different dosage mixes were studied, including the reference mix, with no replacement of waste, and mixes with different percentages of partial replacement of cement with rice husk ash and of natural sand with foundry sand, in separate mixes and also in a single concrete mix. Based on the results obtained, the compressive strength and diametrical compressive tensile strength of the concrete were analyzed, leading to conclusions about the addition of these residues to concrete.

The tests were carried out at the Civil Engineering Laboratory of UNIJUi - Universidade Regional do Noroeste do Estado do Rio Grande do Sul.

3.3 MATERIALS USED

3.3.1 Cement

The type of cement used was CP II-E-32, because it is a cement without pozzolanic additions in its composition and also because it can be used in different types and stages of construction.

Depending on the mixture used, the composition of Portland cement can vary. Table 4 shows the chemical composition of this cement according to InterCement (2015).

Table 4 - Chemical composition of CP II-E-32 cement

CHEMICAL COMPONENTS	CONCENTRATION RANGE
Tricalcium silicate (C3S)	20 - 70
Dicalcium silicate (Ca2Si)	10 - 60
Calcium ferro-aluminate (C4AFe)	5 - 15

Calcium sulphate (CaSO4)	2 - 8
Tricalcium aluminate (C3A)	1 - 15
Calcium carbonate (CaCO3)	0 - 10
Magnesium oxide (MgO)	0 - 6
Calcium oxide	0 - 3

Source: InterCement (2015)

The standards used to carry out the cement physical characterization tests are shown in Table 5, and Figure 14 illustrates some of the test stages. The results are shown in Table 6.

Table 5 - Cement physical characterization tests and their respective standards

Tests	Standards
1) Thinness # 0.075mm	NBR 11579
2) Specific mass	NBR NM 23
3) Start and end of grip	NBR NM 65

Source: author

Figure 14 - Stages of the physical testing of cement

Source: author

Table 6- Characteristics of CP II-E-32 cement

PROPERTY	RESULT
Fineness # 0.075 (%)	0,44
Setting time (hs)	04:46
Specific mass (g/cm $)^3$	3,035

Source: Author

3.3.2 Rice husk ash

The rice husk ash, one of the residues replaced in the concrete, came from the Passo Cereals Industry in the city of Itaqui-RS, identified as "Micro silica MS - 325". It is a microsilica whose base is a finely ground black-gray silica. There was no need to look for this material, as it was already available in the Civil Engineering Laboratory at Unijui.

The characterization of this material was carried out in the Civil Engineering laboratory at Unijui.

The standards used are shown in Table 7 and Figure 15 illustrates some of the test stages. The results obtained are shown in Table 8.

Table 7 - Physical characterization tests for rice husk ash and the respective standards

Tests	Standards
1) Thinness # 0.075mm	NBR 11579
2) Specific mass	NBR NM 23

Source: author

Figure 15 - Stages of the physical testing of rice husk ash microsilica

Source: Author

Table 8 - Characteristics of rice husk ash microsilica

PROPERTY	RESULT
Fineness # 0.075 (%)	0,28
Specific mass (g/cm^3)	1,84

Source: author

Using data obtained from CIENTEC (Science and Technology Foundation), the rice husk ash used in the research had the following characteristics, described in Table 9.

Table 9 - Characteristics of rice husk ash microsilica by CIENTEC

PROPERTY	RESULT
Silica (SiO2) in natura (%)	87,7
Silica (SiO2) in the ash (%)	95,7
Hygroscopic humidity (%)	1,71
Unburned content (%)	8,65
Average diameter (μm)	22,72

Source: CIENTEC (2010)

3.3.3 Medium sand (natural)

The fine aggregate was supplied by the LEC - Civil Engineering Laboratory of Unijui, located in the municipality of Ijui - RS. The physical characterization tests of this material were determined by the standards, as shown in Table 10. Figure 16 illustrates some of the test procedures and Table 11 shows the results.

Table 10 - Physical characterization tests for small aggregates and their respective standards

Tests	Standards
1) Particle size composition	NBR NM 248
2) Specific mass of the fine aggregate	NBR 9776
3) Loose unit mass	NBR NM 45

Source: author

Figure 16 - Procedures for the physical testing of sand

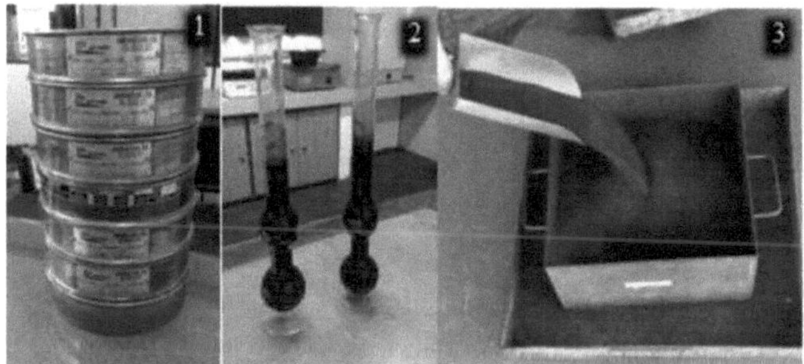

Source: author

Table 11 - characteristics of natural sand

PROPERTY	RESULT
Max. diameter (mm) (mm)	1,2
Fineness modulus	1,9
Specific mass (g/cm $)^3$	2,60
Loose unit mass (kg/dm3)	1,53

Source: Author

3.3.4 Industrial sand

The foundry sand used was donated by the FUNDIMISA company in the municipality of Santo Ângelo - RS, as shown in Figure 17.

For the characterization tests of this material, the same standards were used as for the medium (natural) sand, mentioned above in Table 10, and the same procedures were used to carry out the tests. The results obtained are shown in Table 12.

Figure 17 - Factory specializing in iron casting and machining (Fundimisa)

Source: Fundimisa (2015)

Table 12 - Foundry sand characteristics

PROPERTY	RESULT
Max. diameter (mm) (mm)	0,6
Fineness modulus	1,16
Specific mass (g/cm)^3	2,43
Loose unit mass (kg/dm3)	1,33

Source: Author

3.3.5 Brita 1

The coarse aggregate, gravel 1, as well as the natural sand, was also supplied by Unijui's LEC. The physical characterization tests of this material were determined by the standards shown in Table 13, Figure 18 illustrates some of the test stages, and Table 14 shows the results obtained.

Table 13 - Physical characterization tests for coarse aggregate and their respective standards

Tests	Standards
1) Particle size composition	NBR NM 248
2) Specific mass and absorption of the coarse aggregate	NBR NM 53
3) Compacted unit mass	NBR NM 45

Source: author

Figure 18 - Gravel 1 characterization tests

Source: author

Table 1 - Physical characteristics of coarse aggregate (gravel 1)

PROPERTY	RESULT
Max. diameter (mm)	19
Fineness modulus	6,75
Specific mass (g/cm)^3	2,93
Absorption (%)	1,24
Compact unit mass (kg/dm3)	1,69

Source: Author

3.3.6 Water

The water used to characterize the materials and shape the test specimens came from the artesian well that supplies Unijui's LEC.

3.4 DOSING AND MIXING METHODS

One of the main stages that requires a lot of attention is the dosing process, which is an extremely important activity because, in addition to establishing the quantities of each material that makes up the concrete, it also influences the characteristics responsible for achieving good performance in the final product, related to quality and economy.

Basilio (1977) defines the dosage of concrete as the technique and art of fixing the quantities of its component elements in order to guarantee the characteristics required in the plastic phase and in the phase after hardening.

Once the characterization tests of the concrete's constituent materials had been completed, it was possible to calculate the reference mix and carry out the moulds. To carry out this research, the ABCP Dosage Method was used, which is a method based on tables, calculated in a simple way and the steps to obtain the appropriate proportions of materials were developed.

31

A cone truncated slump of between 80 and 100 mm has been established. To obtain a certain slump, the amount of water depends on the maximum size, shape, texture and granulometry of the coarse aggregate.

After calculating the traits, the amount of material needed to mold 10 specimens per trait was established, 8 of which were used to obtain the compressive strength at ages 7, 14, 21 and 28 days, 2 specimens for each age, and 2 specimens for the diametrical compression tensile strength test, broken at 28 days of age.

After determining the reference mix, specimens were molded with 5%, 10%, 15% and 20% cement partially replaced by rice husk ash microsilica and natural sand partially replaced by foundry sand, first in separate mixtures and then with the two residues in the same mixture.

The molded features were:

-      RF, without the addition of residues;

-      F 5, molded with 5% natural sand replaced by foundry sand;

-      F 10, molded with 10% natural sand replaced by foundry sand;

-      F 15, molded with 15% natural sand replaced by foundry sand;

-      F 20, molded with 20% natural sand replaced by foundry sand;

-      M 5, molded with 5% cement replaced by rice husk ash microsilica;

-      M 10, molded with 10% cement replaced by rice husk ash microsilica;

-      M 15, molded with 15% cement replaced by rice husk ash microsilica;

-      M 20, molded with 20% cement replaced by rice husk ash microsilica;

-      FM 5, molded with 5% replacement of natural sand with foundry sand and 5% replacement of cement with rice husk ash microsilica;

-      FM 10, molded with 10% replacement of natural sand with foundry sand and 10% replacement of cement with rice husk ash microsilica;

-      FM 15, molded with 15% replacement of natural sand with foundry sand and 15% replacement of cement with rice husk ash microsilica;

-      FM 20, molded with 20% replacement of natural sand with foundry sand and 20% replacement of cement with rice husk ash microsilica;

The dosage calculations can be found in Appendix A. Tables 4, 5 and 6 show, respectively, the mixes and quantities of materials for the mix with foundry sand, the mix with rice husk ash microsilica and the mix with the two residues together.

Table 4 - Consumption of materials per trait for replacing natural sand with foundry sand

DRAFT	Cement consumption	Consumption Natural Sand	Consumption of Foundry Sand	Crushed stone consumption	Water consumption	Factor (a/c)
RF	7.76 kg	16.58 kg	X	32.5 kg	5.13 kg	0,66
F 5	7.76 kg	15.75 kg	0.830 kg	32.5 kg	5.13 kg	0,66
F 10	7.76 kg	14.92 kg	1.66 kg	32.5 kg	5.13 kg	0,66
F 15	7.76 kg	14.10 kg	2.49 kg	32.5 kg	5.13 kg	0,66
F 20	7.76 kg	13.26 kg	3.32 kg	32.5 kg	5.13 kg	0,66

Source: author

Table 5 - Material consumption per mix for replacing cement with rice husk ash microsilica

DRAFT	Cement consumption	Consumption Sand Natural	Consumption Rice husk ash microsilica	Crushed stone consumption	Water consumption	Factor (a/c)
RF	7.76 kg	16.58 kg	X	32.5 kg	5.13 kg	0,66
M 5	7.37 kg	16.58 kg	0.390 kg	32.5 kg	5.13 kg	0,66
M 10	6.98 kg	16.58 kg	0.780 kg	32.5 kg	5.13 kg	0,66
M 15	6.60 kg	16.58 kg	1.16 kg	32.5 kg	5.13 kg	0,66
M 20	6.32 kg	16.58 kg	1.55 kg	32.5 kg	5.13 kg	0,66

Source: author

Table 6 - Consumption of materials per design for replacing cement with rice husk ash microsilica and replacing natural sand with foundry sand

DRAFT	Cement consumption	Consumption Sand Natural	Foundry Sand Consumption	Microsilica consumption of rice husk ash	Crushed stone consumption	Water consumption	Factor (a/c)
RF	7.76 kg	16.58 kg	X	X	32.5 kg	5.13 kg	0,66
FM 5	7.37 kg	15.75 kg	0.830 kg	0.390 kg	32.5 kg	5.13 kg	0,66
FM 10	6.98 kg	14.92 kg	1.66 kg	0.780 kg	32.5 kg	5.13 kg	0,66
FM 15	6.60 kg	14.10 kg	2.49 kg	1.16 kg	32.5 kg	5.13 kg	0,66
FM 20	6.21 kg	13.26 kg	3.32 kg	1.55 kg	32.5 kg	5.13 kg	0,66

Source: author

After determining the quantities of materials for each of the mixes, the concrete was produced in a concrete mixer at Unijui's Civil Engineering Laboratory (Figures 19 and 20).

Figure 19 - Materials separated for molding

33

Source: Author

Figure 20 - Materials mixed in a concrete mixer

Source: Author

The slump test was determined to be 80 ±100 mm. These procedures were carried out in accordance with NBR NM 67 (1998). First, the molds and the base plate were moistened, the plate was fixed to the ground with the feet, and the mold was quickly filled with concrete in three layers, each receiving 25 blows from the cone rod. These procedures are illustrated in Figures 21 and 22.

Figure 21 - Performing the Slump Test

Source: Author

Figure 22 - Measuring cone trunk collapse

Source: Author

Once the procedures for determining the cone trunk slump had been completed, the actual water consumption of each mix was corrected, the w/c factor was calculated and the specific mass of the concrete was obtained, as shown in Tables 7, 8 and 9.

Table 7 - Slump test values and final w/c factor for replacing natural sand with foundry sand in concrete

DRAFT	Rebate mm	Consumption real water (l)	a/c factor corrected	Specific Weight of Concrete (kg/m^3)

	90	4,380	0,56	22,84
TR	90	4,380	0,56	22,84
F 5	85	3,310	0,55	22,93
F 10	100	4,480	0,58	22,37
F 15	90	4,440	0,57	22,80
F 20	90	4,610	0,59	22,62

Source: author

Table 8 - Slump test values and final w/c factor for replacing cement with rice husk ash microsilica in concrete

DRAFT	Rebate mm	Consumption real water (l)	a/c factor corrected	Specific Weight of Concrete $(kg/m)^3$
TR	90	4,380	0,56	22,84
M 5	90	4,360	0,59	22,80
M 10	90	4,420	0,63	22,75
M 15	85	4,480	0,68	22,86
M 20	95	4,640	0,73	22,80

Source: author

Table 9 - Slump test values and final w/c factor for replacing natural sand with foundry sand and cement with rice husk ash microsilica in concrete

DRAFT	Rebate mm	Actual water consumption	Corrected a/c factor	Specific Weight of Concrete
TR	90	4,380	0,56	22,84
FM 5	85	4,390	0,59	23,17
FM 10	95	4,640	0,66	22,85
FM 15	95	4,770	0,72	22,56
FM 20	90	4,810	0,77	22,54

Source: author

The specimens were then molded in accordance with standard NBR 5738 (2003), and the dimensions of the molds used were 10 x 20 cm. The molds were first coated with a thin layer of mineral oil, then 2 layers of concrete were placed in each one, and each layer was compacted 12 times (Figure 23).

Figure 23 - Molding the specimens

Source: author

After molding, the specimens were exposed to room temperature for 24 hours, then demolded and taken to a humid chamber at a temperature of 23 ± 2 °C and humidity > 95%, where they remained until the test dates, which were at 7, 14, 21 and 28 days of age for the compressive strength test and 28 days for the diametrical compressive tensile strength test. Figures 24, 25 and 26 illustrate the procedures described.

Figure 24 - Specimens exposed to room temperature

Source: author

Figure 25 - Demoulded, measured and numbered specimens to be taken to the wet chamber

Figure 26 - Specimens positioned in the wet chamber

3.4.1 Compressive strength test

The axial compressive strength of the concrete was assessed using cylindrical specimens measuring 10 cm in diameter and 20 cm in height. This test was carried out at Unijui's Civil Engineering Laboratory. Using a standardized press, the strength of the concrete was determined at ages 7, 14, 21 and 28 days from the dates on which the specimens were cast. The procedures were carried out in accordance with NBR 5739 (2007), and the specimen had to be centered in the press so that its axis was aligned with that of the press, so that the resultant forces passed through the center (Figure 27). The compressive strength was calculated according to Equation (1) below:

$$fc = \frac{4F}{\pi x D^2}$$

Equation (1) Where:

FC: is the compressive strength, in MPa;

F: is the maximum force achieved, in Newtons;

π: Pi;

D: is the diameter of the specimen in mm.

Figure 27 - Compressive Strength Tests

Source: Author

3.4.2 Diametral compression tensile strength test

The diametrical compression tensile strength was analyzed using two specimens for each trait, measuring 10 cm in diameter and 20 cm in height, 28 days after casting. Carried out at Unijui's LEC, the specimen was positioned at rest along a generatrix on the plate of the compression machine, as described by NBR 7222 (2011). The load was applied continuously, without shock, with a constant increase in tensile stress, at a speed of 0.05 + 0.02 MPa/s until the specimen ruptured (Figure 28). The tensile strength by diametrical compression was determined using Equation (2):

$$F_{t,D} = \frac{2F}{\pi dL}$$

Equation (2)

$F_{t,D}$ = tensile strength by diametrical compression, expressed in MPa, to the nearest 0.05 MPa;

F = maximum load obtained in the test, in kN;

d = diameter of the specimen, in mm;

L = height of the specimen, in mm.

Figure 28 - Diametral compression tensile strength tests

Source: Author

3.4.3 Scanning Electron Microscopy test

Studies were carried out on this test, with the intention of carrying it out, but as the institution does not have the necessary equipment and the university with which this type of test used to be carried out was on strike, it was unfortunately not possible to carry it out.

4 ANALYSIS AND DISCUSSION OF RESULTS

This section will present the results obtained from the Compressive Strength and Diametral Compressive Tensile Strength tests on Portland cement concrete.

4.1 COMPRESSION RESISTANCE TEST - CONCRETE WITH PARTIAL REPLACEMENT OF NATURAL SAND WITH FOUNDATION SAND

The compressive strength test yielded the results shown in Figure 29. It can be seen that the reference mix (RF), molded without the addition of waste, achieved a higher compressive strength at 7 days of age compared to the other mixes molded with the replacement of natural sand with foundry sand. However, this reduction in strength is significant, since the 5% and 10% replacements resulted in strengths very close to the reference mix (RF).

The results found differ greatly from those presented by Lima (2014), in which by replacing percentages of natural sand with foundry sand, from 5% up to 50%, the aforementioned author obtained higher strengths than the reference design for all ages. This can be explained by the type of cement used in the molding of the specimens, with the aforementioned author using CP IV - 32, and in the present study CP II E - 32 was used, since the materials that make up the cement influence the properties of the concrete, and especially the compressive strength.

Figure 29 - Compressive strength of concrete at 7 days of age

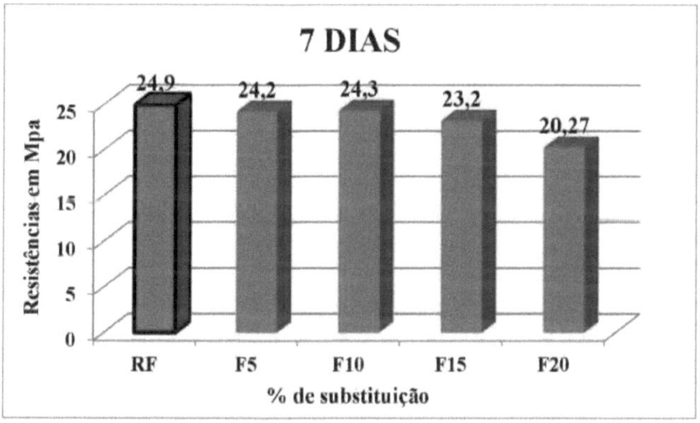

Source: Author

When looking at the compressive strength of the concrete at 14 days of age (Figure 30), it can be seen that the reference mix had the highest strength among the others, but the mix with 5% replacement of natural sand with foundry sand (F5) remained close, reaching a strength of 30.3 Mpa, 1.5 Mpa lower than the reference mix (RF).

Figure 30 - Compressive strength of concrete at 14 days of age

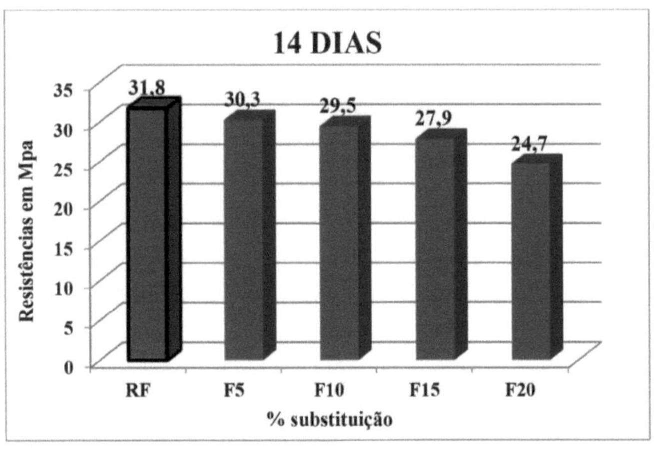

Figure 30 - Compressive strength of concrete at 14 days of age

Source: Author

Figure 31 shows the evolution of the compressive strengths at 21 days of age. The reference mix (RF) reached 33.9 Mpa, while the other mixes remained at lower strengths, except for the mix with 10% replacement of natural sand with foundry sand (F10), which reached 34.2 Mpa.

Figura 31   - Compressive strength of concrete at 21 days of age

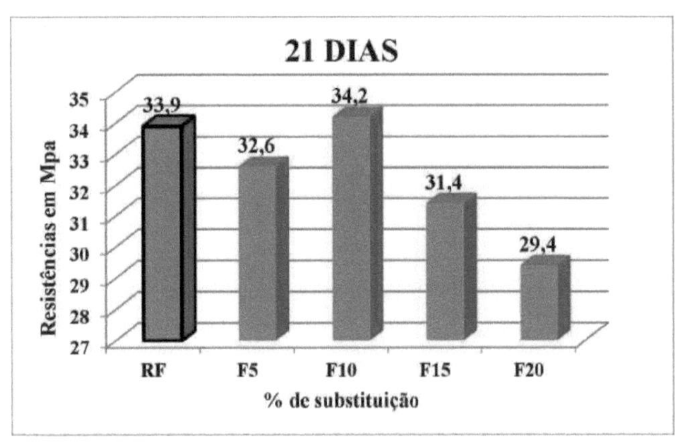

Source: Author

At 28 days of age, with a longer curing time for the concrete, two mixes had higher strength values than the reference mix (RF), which remained at 33.5 Mpa, with the mix with 5% replacement of natural sand with foundry sand (F5) having 36.4 Mpa and the mix with 10% replacement of natural sand with foundry sand (F10) having a compressive strength of 34.1 Mpa, while the other mixes remained constant (Figure 32). The increase in strength with the use of foundry sand can be

42

explained by the sand's characterization tests, with the fineness modulus of the natural sand resulting in 1.9 and the fineness modulus of the foundry sand 1.16. Because foundry sand has finer grains than natural sand, it can better fill the voids in the concrete and also contribute to better hydration.

Figura 32   - Compressive strength of concrete at 28 days of age

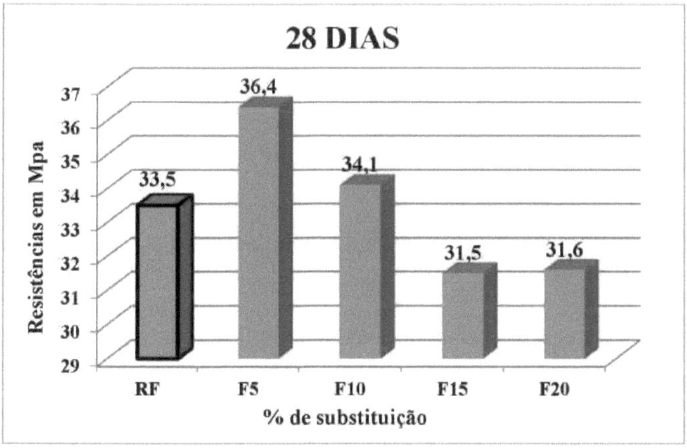

Source: Author

When the concrete was 28 days old, Lima (2014) mentioned in his research that all the mixes had higher compressive strengths than the reference mix, and the highest strength was achieved with a 5% substitution, which was 8.17 MPa higher than the result found in the reference mix. In the research under study, however, there was an increase in strength, but only two types of concrete resulted in values higher than the reference concrete (RF).

32.1.1 Percentage presentation of the evolution of compressive strengths of concrete with foundry sand

Table 15 shows the percentage change in compressive strength of each concrete mix during the period from 7 days to 28 days.

Table 15 - Evolution of compressive strength in percentages

Dash	Compressive strength (MPa)		Increase in Resistance (%)
	7 days	28 days	
RF	24,9	33,5	35
F5	24,2	36,4	50
F10	24,3	34,1	40

| F15 | 23,2 | 31,5 | 36 |
| F20 | 20,27 | 31,6 | 56 |

Table 15 shows that the mix with the highest compressive strength evolution during the 7-day period when the concrete was 28 days old was the mix with 20% natural sand replaced by foundry sand (F20), reaching an evolution of 56%. The second highest evolution occurred with the F5 mix, which reached 50% during the same period. The reference steel (RF) had a resistance evolution of 35% and the F10 and F15 steel reached values close to this, with 40% and 36% respectively. The greatest strength increases over the same period occurred in the mixes in which foundry sand was substituted, which may be explained by the existence of some of the chemical components that make up this sand.

4.2 COMPRESSIVE STRENGTH TEST - CONCRETE WITH PARTIAL CEMENT REPLACEMENT BY RICE HUSK ASH

Figure 33 shows the results of the 7-day compressive strength test on the concrete. It shows that the reference mix (RF) achieved a higher strength than the others, with the strengths decreasing as the cement replaced by rice husk ash increased. The difference in strength between the reference mix (RF) and the mix with the lowest strength, with 20% replacement (M20), was 12 Mpa, which is quite significant.

Kelm (2013) obtained lower compressive strengths at 7 days of age for all the concrete mixes compared to the reference mix, and found that the higher the percentage of replacement, the lower the compressive strength. These results coincide with the present study.

Figure 33 - Compressive strength of concrete at 7 days of age

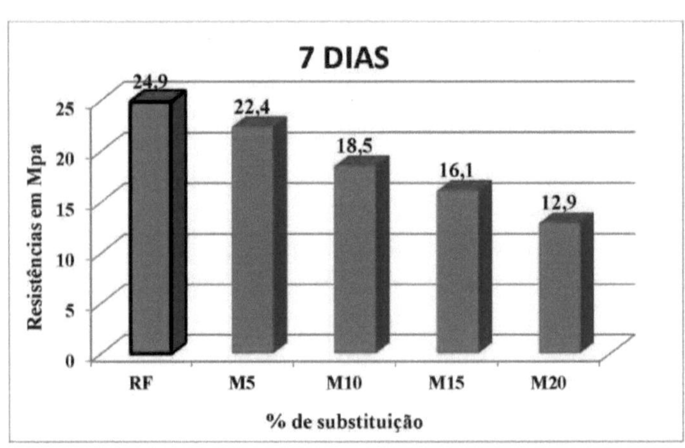

When the concrete was 14 days old, Figure 34 shows that the reference mix (RF) continued to be the mix with the highest strength, reaching 31.8 Mpa, while the mix with 5% cement replaced by rice husk ash was close, reaching 29.6 Mpa, and the other mixes decreased progressively as the replacement content increased, reaching 19.1 Mpa with the 20% replacement (M20).

Figura 34   - Compressive strength of concrete at 14 days of age

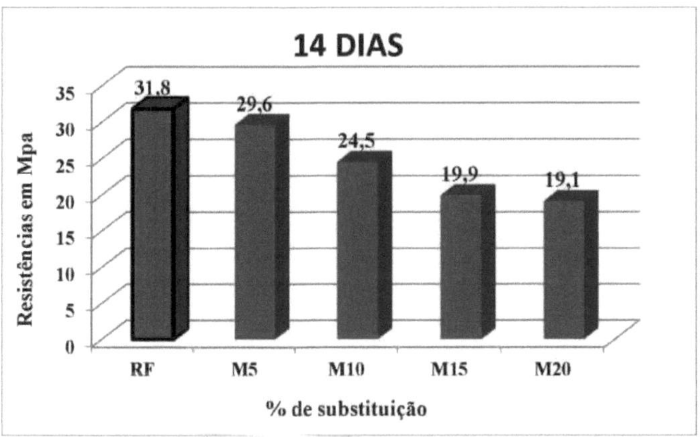

As the days went by, with a longer curing time, the strength of the concrete increased, and even so the reference mix (RF) at 21 days of age continued to be higher than the mixes with cement replaced by rice husk ash (Figure 35). Even with the reduction in strength due to the substitution, the mix with the highest percentage of substitution reached a compressive strength of 21.5 Mpa, which is a very considerable strength for various types of works and also an economical concrete mix, as cement is the most expensive input used in concrete production, and this mix reduces the amount of cement used in the reference mix (RF) by 20%.

Figura 35   - Compressive strength of concrete at 21 days of age

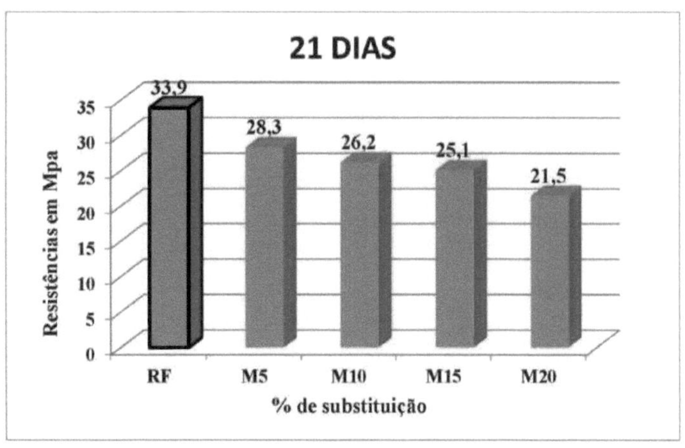

Source: Author

When the concrete was 28 days old, the reference mix (RF) was the one that achieved the highest strength (Figure 36), but the others did not differ much, with the mix with the highest percentage of substitution (M20) achieving a compressive strength of 25.8 Mpa, a very significant value due to the high percentage of rice husk ash that was added to the concrete.

Figure 36 - Compressive strength of concrete at 28 days of age

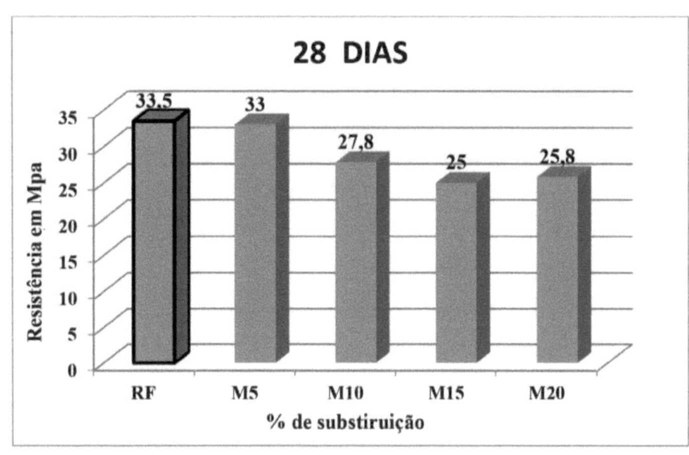

Source: Author

Kelm (2013) pointed out in his studies that the hydration reactions of the cement components and also the pozzolanic reaction process of the mixtures with rice husk ash microsilica had progressed by 28 days, resulting in compressive strengths of the specimens very close to the reference mixture, where the difference between the reference mixture and the highest substitution content was around

6.34 Mpa.

4.2.1 Percentage presentation of the compressive strength evolution of concrete with rice husk ash

Table 16 shows the evolution of the concrete compressive strengths between the ages of 7 and 28 days.

Table 16 - Evolution of compressive strength in percentages

Dash	Compressive strength (MPa)		Increase in Resistance (%)
	7 days	28 days	
RF	24,9	33,5	35
M5	22,4	33	47
M10	18,5	27,8	50
M15	16,1	25	55
M20	12,9	25,8	100

Source: Own authorship

Table 16 shows that the greatest increase in compressive strength between the ages of 7 and 28 days occurred with the mix in which 20% of the cement was replaced by rice husk ash (M20), with a 65% increase in strength compared to the change in strength achieved by the mix molded without the addition of waste, for the same period. We can also see that the strength evolution for each of the traits, analyzed over the same period (7 to 28 days), was proportional to the increase in the percentage of substitution, where the M5 trait showed a strength evolution of 47% and the other traits, M10 and M15 showed a strength evolution of 50% and 55% respectively. This increase in compressive strength in the traces with cement replaced by rice husk ash may have been caused by the different shape of the rice husk ash grains, and explained by the characterization tests of these materials in which the cement used had a specific mass of 3.035 g/cm^3 and the rice husk ash 1.84 g/cm^3 , a very significant difference which may probably have interfered with the results.

4.3 COMPRESSIVE STRENGTH TEST - CONCRETE WITH PARTIAL REPLACEMENT OF NATURAL SAND WITH FOUNDRY SAND AND WITH PARTIAL REPLACEMENT OF CEMENT WITH RICE HUSK ASH

Figure 37 shows the results obtained by replacing the two residues in the same concrete mix, i.e. partially replacing natural sand with foundry sand and cement with rice husk ash. As can be seen in Figure 37, the reference mix (RF) achieved higher strengths than the others, with the strengths decreasing as the replacement levels increased, except for the FM15 mix, which had a small increase of 0.6 Mpa compared to the FM10 mix.

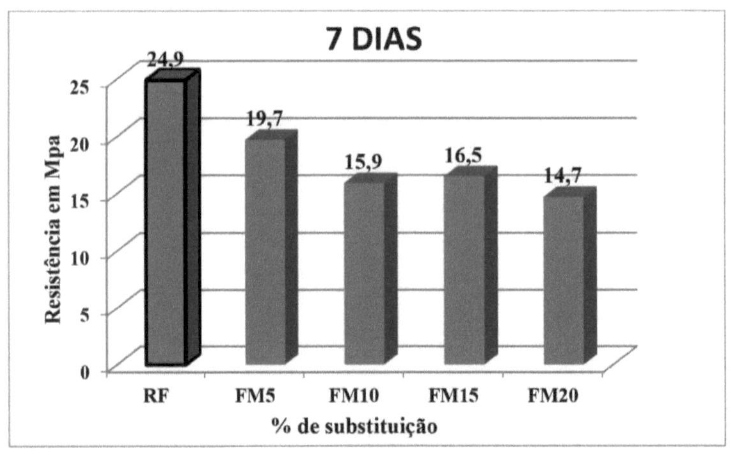

Source: Author

At 14 days of age, the mix with the highest compressive strength was still the reference mix (RF), as can be seen in Figure 38. The mix with 5% natural sand replaced by foundry sand and 5% cement replaced by rice husk ash obtained a strength very close to the result found in the reference mix, the difference between the two being 2.9 MPa.

Figure 38 - Compressive strength of concrete at 7 days of age

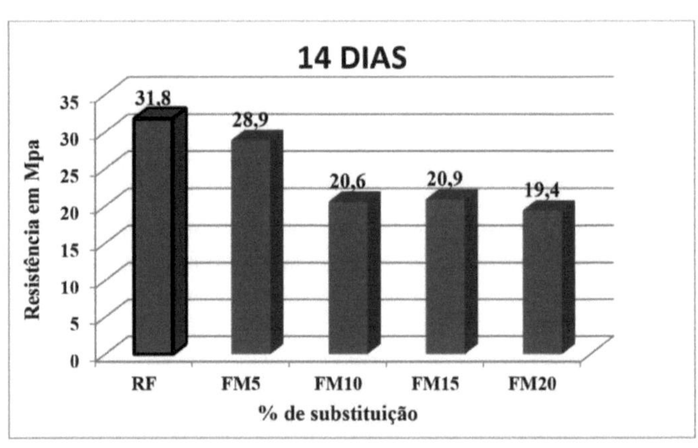

Source: Author

When the concrete was 21 days old, the reference mix (RF) had a higher compressive strength than the other mixes, as shown in Figure 39. As can be seen, the strengths decrease as the percentage of substitution increases, except for the 20 % substitution (FM 20), which achieved a higher strength than the FM 15 mix.

Figura 39   - Compressive strength of concrete at 21 days of age

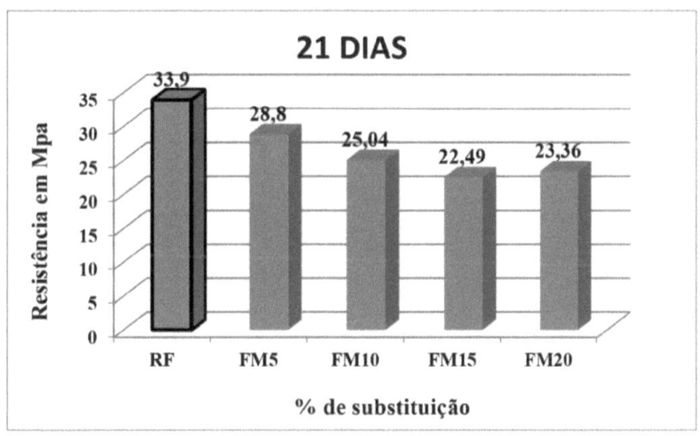

Source: Author

Figura 40   Source: Author

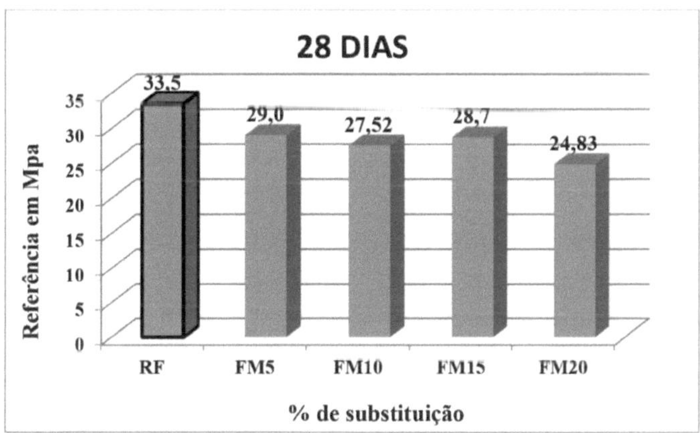

Source: Author

Figure 40 shows that the mix with the highest compressive strength was the reference mix (RF), which reached 33.5 Mpa at 28 days of age. The strength of the other mixes decreased as the percentage of replacement increased, with FM15 having an increase of 1.18 Mpa compared to FM10. The mix with the highest percentage of replacement (FM 20) obtained 24.83 Mpa, 8.67 Mpa less than the reference mix (RF).

Figura 41   - Compressive strength of concrete at 28 days of age

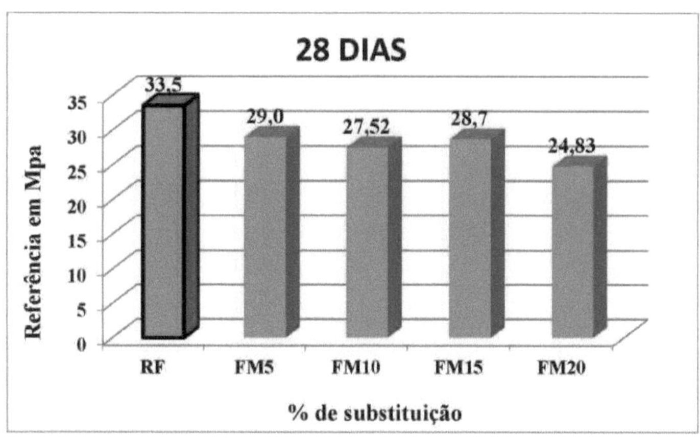

Source: Author

40.3.1 Percentage presentation of the evolution of compressive strengths of concrete with foundation sand and rice husk ash

Table 17 shows the evolution of the compressive strengths of each of the mixes between the ages of 7 and 28 days.

When analyzing Table 17 above, it can be concluded that there was an increase in strength proportional to the replacement of waste during the same period (from 7 to 28 days), except for the mix with 20% replacement of natural sand with foundry sand and 20% replacement of cement with rice husk ash (FM20) which resulted in an increase of 69%, 5% less than the mix (FM15) which achieved 74% increase. The reference course showed a resistance evolution of 35% over the same period. The FM5 and FM10 traits, on the other hand, evolved by 47% and 73% respectively. This increase in compressive strength may have been due to the different characteristics and compounds of the waste that was replaced in the mixtures, as well as the percentage of replacement in each mix.

Table 17 - Evolution of compressive strength in percentages

Dash	Compressive strength (MPa)		Increase in Resistance (%)
	7 days	28 days	
RF	24,9	33,5	35
FM5	19,7	29	47
FM10	15,9	27,5	73
FM15	16,5	28,7	74
FM20	14,7	24,8	69

Source: Author

4.4 COMPRESSIVE STRENGTH COMPARISONS

Figure 41 shows all the mixes and their strengths at 7 days of age, as can be seen below. We can see that the highest strength was achieved by the mix with no replacement of waste (RF), with a compressive strength of 24.9 MPa. The other mixes in which natural sand was replaced by foundry sand (F) achieved higher strengths than the mixes in which cement was replaced by rice husk ash (M). On the other hand, the mixes with natural sand replaced by foundry sand and cement replaced by rice husk ash (FM) achieved lower strengths than the mixes molded with only one of the residues replaced, except for the FM15 and FM20 mixes, which resulted in 16.5 Mpa and 14.7 Mpa respectively in compressive strength, higher than the results found in the M15 and M20 mixes.

Figura 41   - Compressive strength of concrete of the three types of mixtures at 7 days of age

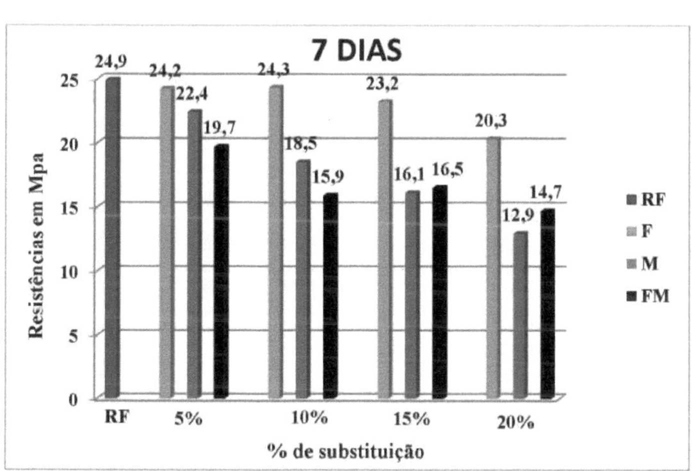

Source: Author

From Figure 41 analyzed above, it can be concluded that the residues that were replaced in the concrete interfere with its properties. Due to the replacement of these residues in the concrete, the compressive strengths varied, which could be caused by the chemical composition of these residues, the granulometric properties of each of the residues, and even by errors during the molding of some traces.

Figure 42 shows that all the mixes had lower compressive strengths than the reference mix (RF), which reached 33.5 Mpa at 28 days of age, except for mixes F5 and F10, which reached 36.4 Mpa and 34.1 Mpa, respectively. In the other mixes with rice husk ash (M) and with the two residues in the same mix (FM), it can be seen that the results were close and that all the mixes with only rice husk ash resulted in higher compressive strength than the mixes with the two residues, except for mix FM15, which resulted in 28.7 Mpa, 3.7 Mpa higher than mix M15.

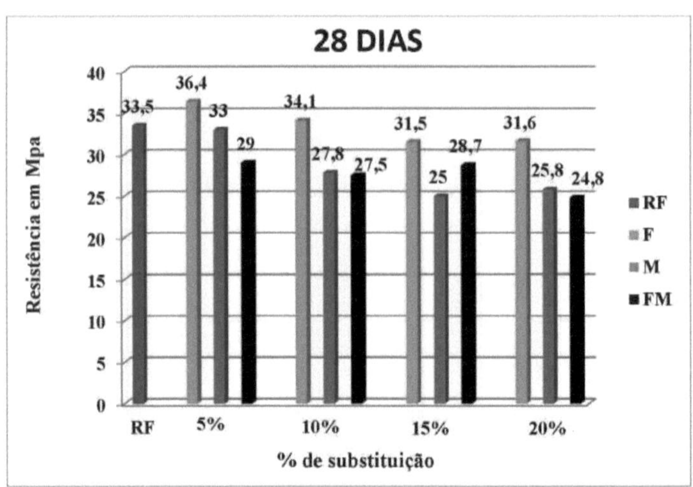

Source: Author

By analyzing Figure 42, we can see that the compressive strengths of the concrete varied greatly, and the increase in strength with the replacement of natural sand with foundry sand can perhaps be explained by its granulometric composition, with the fineness modulus of the natural sand being 1.9 and the fineness modulus of the foundry sand being 1.16, so the foundry sand may have filled the voids in the concrete better and hydration occurred better, thus increasing the strength.

The traces with only cement replaced by rice husk ash resulted in lower strengths than the reference trace, which can perhaps also be explained by the granulometric composition of this residue, in which the cement used had a specific mass of 3.035 g/cm^3 and the rice husk ash 1.84 g/cm^3 , a very significant difference that probably interfered with the results, since the cement was partially replaced by the CCA.

And the mixes with the two residues in the same mix did not differ much from the mixes with only CCA. It can thus be seen that the residue that most interfered with the strength of the concrete was CCA, and these results can perhaps be explained by the fact that cement is the material it replaces and that it is primarily responsible for the strength of the concrete.

4.5 DIAMETRICAL COMPRESSIVE TENSILE STRENGTH OF CONCRETE

The results illustrated in Figure 43 show that the reference mix (RF) achieved a tensile strength of 3.23 MPa, which is lower than most of the results found in the mixes with waste substitution. All the mixes molded only with the replacement of natural sand with foundry sand (F) obtained higher strengths than the reference mix (RF), where the highest strength achieved was 4.39 Mpa, obtained by the mix with 10% replacement of natural sand with foundry sand (F10). The strengths of the mixes molded only with cement replaced by rice husk ash were close to those of the mixes with only foundry sand, but lower than them. The mix that came closest was M5, with a strength of 4.19Mpa. The mixes in which the two residues were replaced had lower strengths than the mixes molded with the replacement of just one of them. The lowest strength found was in mix FM15, with a strength of 2.88Mpa, and the highest strength was in mix FM10, with 3.55Mpa.

Figure 43 - Diametral compressive tensile strength of concrete of the three types of mixtures at 28 days of age

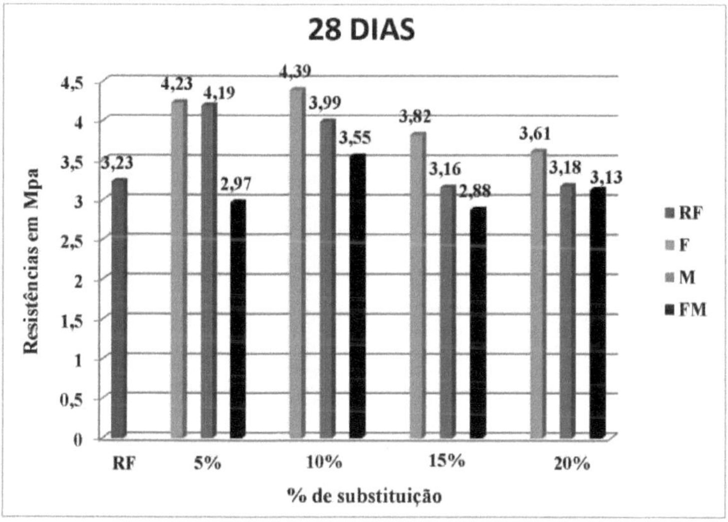

Source: Author

In relation to the compressive strengths, it can be seen from Figure 43 that for the same traits analyzed, the diametrical compressive tensile strength results in higher strength values in almost all traits with the addition of residues compared to the reference trait.

The results shown in Figure 43 may have been influenced by the chemical composition of the waste, as well as by some errors during the tests.

5   CONCLUSION

The aim of this study was to analyze the compressive strength and tensile strength by diametrical compression of concrete, by partially replacing cement with rice husk ash microsilica and natural sand with foundry sand, at different replacement levels, 5%, 10%, 15% and 20%, first in separate mixes and then with the two residues in the same concrete mix.

However, the different concrete properties were analyzed using the following topics:

5.1   COMPRESSIVE STRENGTH

By partially replacing natural sand with foundry sand, analyzed at all replacement levels, 5%, 10%, 15% and 20% and at the different ages of the concrete, it can be seen that the traces cast without the addition of this residue resulted in higher strengths at almost all ages. However, at 21 days of age, the 10% substitution resulted in a higher strength than the reference mix, while the other substitutions remained close but lower. At the age of 28 days, the 5% and 10% substitutions obtained higher results than the reference mix. Even with the increase in the percentage of natural sand replaced by foundry sand, the mixes obtained strengths close to the reference mix.

With the replacement of cement with rice husk ash, it was noticed that all the mixes molded with the different replacements resulted in lower strengths than the reference mix, molded without the addition of waste, and the higher the percentage of replacement, the lower the strength found. At the age of 28 days, with a longer curing time for the concrete, the results obtained were very close to the reference mix.

In the traces molded with the two residues in the same mixture, foundry sand and rice husk ash, in the substitutions of 5%, 10%, 15% and 20% and at the ages of 7, 14, 21 and 28 days of the concrete, the results found were lower than the reference trace, but with a longer curing time, at 21 and 28 days of age of the concrete, the results were able to reach values closer to the reference trace.

However, it can be concluded that the addition of these residues to concrete is feasible and can be recognized by the construction market, as many designs exceed the strength obtained by the design without the addition of residues. In addition to increasing resistance and improving the properties of concrete, it also reduces the consumption of raw materials and consequently contributes to minimizing environmental impact.

5.2   TENSILE STRENGTH BY DIAMETRICAL COMPRESSION

The diametrical compressive tensile strength tests were carried out on the concrete at 28 days of age and with the different percentages of substitutions, 5%, 10%, 15% and 20%, first in separate mixes and then with the two residues in the same concrete mix.

The concrete mixes in which only natural sand was replaced by foundry sand obtained very satisfactory results, as all the mixes resulted in higher strengths than the reference mix.

When only cement was substituted for rice husk ash, the results obtained remained close to the reference mix, with the 5% and 10% substitutions achieving values higher than the reference mix, and the other mixes obtaining results that were close to, but lower than, the reference mix.

When the two residues were replaced in the same mix, foundry sand and rice husk ash microsilica, all the results obtained were lower than the reference mix, molded without the addition of residues, except for the 10% replacement mix, which achieved higher strength than the reference mix.

When these tests were carried out, it was noticed that some of the traces varied a lot, which may have been caused by errors during the test. But even so, it can be seen that the introduction of these residues into concrete is feasible, being an excellent alternative for reusing these residues and contributing to improving the properties of Portland cement concrete.

However, it is worth pointing out that for these concrete mixes to be produced on a large scale, further studies are needed, especially in relation to the chemical composition of the waste, as well as the microstructure of the concrete, which may shed more light on the results obtained.

5.3 SUGGESTIONS FOR FUTURE WORK

- Study with other types of cement;

- Production of concrete with the same traits on a large scale;

- Analysis of the chemical composition of the constituent materials of concrete.

6 REFERENCES:

ABIFA: **Brazilian Foundry Association**. Available at:

<http://www.abifa.org.br>. Accessed on: March 30, 2015.

Albrecht Equipamentos Industriais: **Regeneration of chemically bonded sands using an environmentally friendly process.** Available at: <http://www.albrecht.com.br/pt-br/default.php?go=nov-5-7-06-areias>. Accessed on: March 12, 2015.

ALMEIDA; DA LUZ (Ed), **Manual de Agregados para Construçâo Civil.** Rio de Janeiro, 2009. 245p.

BRAZILIAN PORTLAND CEMENT ASSOCIATION. **Basic guide to the use of portland cement**. 7.ed. Sao Paulo, 2002. 28p.

NATIONAL ASSOCIATION OF AGGREGATE PRODUCERS FOR CIVIL CONSTRUCTION - ANEPAC. ANEPAC Yearbook 2011. Sao Paulo, ANEPAC, 2011.

ASSOCIAÇÂO BRASILEIRA DE NORMAS TÉCNICAS **NBR 10004**: Residuos sólidos - Classificaçao. Rio de Janeiro, 2004.

. **NBR 11578:** Composite Portland cement - Specification. 1997. 5p.

. **NBR 11579:** Portland cement - determination of fineness using the 200 sieve.

. **NBR 7211:** Aggregates for concrete - specifications. Rio de Janeiro, 2009. 7p.

. **NBR NM 23:** Portland cement and other powdered materials: determination of specific mass. Rio de Janeiro, 1998. 4p.

. **NBR NM 65:** Portland cement - determination of setting time. Rio de Janeiro, 2002. 6p.

. **NBR NM 248:** Aggregates - determination of granulometric composition. Rio de Janeiro, 2001. 5p.

. **NBR 9776:** Aggregates - Determination of the specific mass of fine aggregates using the Chapman flask. Rio de Janeiro, 2003. 8p.

.**NBR 9779:** Hardened mortar and concrete - Determination of water absorption by capillarity. 2012. 3p.

.**NBR NM 45:** Aggregates - Determination of unit mass and void volume. 2006. 8p.

. **NBR NM 53:** Coarse aggregate - determination of specific mass, apparent specific mass and water absorption. Rio de Janeiro. 2002; 4p.

. **NBR NM 67:** Concrete - determination of consistency by cone trunk slump. Rio de Janeiro, 1998. 8p.

. **NBR 5738**: Molding and curing of cylindrical concrete specimens. Rio de Janeiro, 2003.

. **NBR 5732**: Ordinary Portland cement - Specification. Rio de Janeiro, 1991

. **NBR 5739:** Compression tests on cylindrical specimens. Rio de Janeiro, 2007. 9p.

. **NBR 7222:** Concrete and mortar - Determination of tensile strength by diametrical compression of cylindrical specimens. 2011. 5 p.

. **NBR 12653:** Pozzolanic materials - Specification. 2015. 3p.

BASILIO, F. A. **General Considerations on Concrete Batching.** In: COLLOQUIUM ON CONCRETE DOSAGE, May 1977, Sao Paulo. Sao Paulo: IBRACON, 1977.

BAUER, Luiz Alfredo Falcao (Org), **Materiais de construção,** 5°ed. Rio de Janeiro: Editora LTC, 2000, V1, 471p.

BRIZOLA, Rodrigo Matzenbacher. **Microstructure of concrete coverings with high.**

slag and fly ash content activated by portland cement and hydrated lime. 2007. 179p. Dissertation (Master's Degree) - Federal University of Santa Maria, 2007.

COLLINS, L. and FOX, R. A. (1985). **Aggregates: sand, gravel and crushed rock aggregates for construction purposes. The Geological Society publ.**, 220 p.

CONAB - National Supply Company - **Market Prospecting Studies**. Available at

<http://www.conab.gov.br/OlalaCMS/uploads/arquivos/12_09_11_16_41_03_prospeccao_12_13.pdf> . Acesso em: 8 abril 2015.

DANTAS, José. **Assembly, Commissioning and Operation of a Foundry Sand Recovery System**: Thermal Regenerator - Phase II Work Plan. Sao Paulo: Nuclear and Energy Research Institute, 2003.

DUART, Marcelo Adriano. **Study of the microstructure of concrete with the addition of unprocessed residual rice husk ash.** 2008. 134p. Dissertation (Master's Degree) Federal University of Santa Maria, 2008.

FIHP - **Federación Iberoamericana de Hormigón Premesclado.** Available at

<http://www.hormigonfihp.org/>. Accessed on: March 20, 2015.

FRIZZO, Benildo Tocchetto. **Influence of pozzolan content and fineness on oxygen permeability and capillary absorption of concrete.** 2001. 158p. Dissertation (Master's Degree) -

Federal University of Santa Maria, 2001.

STATE FOUNDATION FOR ENVIRONMENTAL PROTECTION. **Report on the generation of industrial solid waste in the state of Rio Grande do Sul**, Rio Grande do Sul, 2003.

FUNDIMISA - **Fundiçao e Usinagem Ltda**. Available at http://www.fundimisa.com.br/>. Accessed on: March 12, 2015.

GONÇALVES, Gislayne Elisana et al. **Synthesis and characterization of mullite using silica obtained from rice husks.** Rem: Ver. Esc. Minas, Ouro Preto, V, Sep 2009.

GOSSEN, Marcell. **Industrial solid waste management program:** Proposal for a procedure and application. 2005. 154 f. Dissertation (Master's in Environmental Engineering) - Regional University of Blumenau Center for Technological Sciences - Blumenau, 2005.

GUEDERT, L. O; **Study of the technical and economic viability of using rice husk ash as a pozzolanic material**. Master's dissertation, PEPS UFSC 1989, 147 p.

HELENE, Paulo; TERZIAN, Paulo. **Manual de dosagem e Controle do Concreto.** Sao Paulo: Pini; 1992. 349 p.

HOUSTON, D. F. **Rice Hulls**. Rice Chemistry and Technology, American Association of Cereal Chemists, p. 301-340, MN, 1972.

INTERCEMENT, Brazil. **Building Sustainable Partnerships.** Available at: http://www.intercement.com.br/ProdutosServicos/cimento. Accessed on: March 29, 2015.

BRAZILIAN CONCRETE INSTITUTE - IBRACON. **Concrete: the most consumed building material in the world. Revista Concreto e Construçôes.** Sao Paulo, v . XXXVII, n. 53, p. 1   -80, Mar. 2009. Available at:

<http://ibracon.org.br/publicacoes/revistas_ibracon/rev_construcao/pdf/Revista_Concreto_53. pdf >. Acesso em: 4 março 2015.

IRGA. Instituto Riograndense do Arroz. **State Agricultural Research Foundation,** July 2, 2012. Available em:<www.fepagro.rs.gov.br/.../20120702100451revista_pag_v17_n1_on>. Acesso em: 20 agosto 2015.

ISAIA, G. C. (Ed.). **CONCRETE: Teaching, Research and Achievements.** Sao Paulo: IBRACON, 2005, v-1.

ISAIA, G. C. **Effect of binary and ternary mixtures of pozzolan in high performance concrete:**

a study of durability with a view to reinforcement corrosion. 1995. Thesis (Doctorate) - Polytechnic School of the University of Sao Paulo, Sao Paulo, 1995.

KELM, Tamile Antunes. **Strength and microstructure analysis of concretes with partial replacement of cement by rice husk ash microsilica.** 2011. 56 f. Course Conclusion Paper (Graduation in Civil Engineering) - Universidade Regional do Noroeste do Estado do Rio Grande do Sul, Ijui, 2011.

LIMA, G. T.S. **Strength and microstructure analysis of concrete with partial replacement of natural sand by foundry sand.** 2014. Course Conclusion Paper. Civil Engineering Course, Universidade Regional do Noroeste do Estado do Rio Grande do Sul - UNIJUi, Ijui, 2014.

LOPES, Luiz Rogério Natividade. **Evaluation of the reduction of solid residues of resin sand in steel castings through thermal recovery.** 2009. 94 f. Dissertation (Master's in Industrial Engineering) - Federal University of Bahia, Salvador, 2009.

LUDWING, D.G. **Concrete with the addition of rice husk ash.** 2014. Course Conclusion Paper. Civil Engineering Course, Centro Universitàrio UNIVATES, Lajeado, 2014.

MATOS, S. V.; SCHALCH, V. **Alternatives for minimizing waste from the foundry industry.** BRAZILIAN CONGRESS OF SANITARY AND ENVIRONMENTAL ENGINEERING, 1997.

MEHTA, P. K. in: **Proceedings of II Intern. Conference on High-performance Concrete and...,** Gramado, RS, 1999.

MEHTA, P. K; MONTEiRo, P. J. M. **Concreto: Estrutura, Propriedades e Materiais.** Sao Paulo: PiNi, 1994.

MEHTA, P. K.; MONTEiRo, J. P. M. **Concreto: microestrutura, propriedades e materiais.** Sao Paulo: Brazilian Concrete Institute, 2008.

MILANI, A. P. S. **Physical, Mechanical and Thermal Evaluation of the Soil-Cement-Rice Husk Ash Material and its Performance as a Monolithic Wall.** PhD thesis. State University of Campinas, Faculty of Agricultural Engineering. Campinas - SP, 2008.

MISSAU, Fabiano. **Chloride penetration in concrete containing different levels of rice husk ash.** 2004. 146 p. Master's degree dissertation - Federal University of Santa Maria, 2004.

NEVILLE, Adam Matthew. **Properties of concrete.** Sao Paulo: Pini, 1997. 150 p.

PEiXoTo, F. **Thermal regeneration of chemically bonded sand.** Master's dissertation. Santa Catarina State University, Joinville, 2003.

PoUEY, M. T. F. **Processing of residual rice husk ash with a view to producing composite**

and/or pozzolanic cement. 2006. Thesis (Doctorate in Civil Engineering) - Federal University of Rio Grande do Sul, Porto Alegre, BRRS, 2006.

SALGADO, J. **Técnicas e prâticas construtivas para edificaçâo**. 2. ed. Sao Paulo: Èrica, 2009.

SANTOS, S. (1997). **Feasibility study on the use of residual rice husk ash in mortars and concretes**. Dissertation (Master's Degree) - Federal University of Santa Catarina, 1997.

SCHEUNEMANN, Ricardo. **Regeneration of foundry sand through chemical treatment via the fenton process.** Master's dissertation - Chemical Engineering Course - Federal University of Santa Catarina - Florianópolis, 2005.

SIDDIQUE, R.; SINGH, G. **Utilization of waste foundry sand (WFS) in concrete manufacturing.** Resources, Conservation and Recycling, Volume 55, Issue 11, September 2011, Pages 885-892, ISSN 0921-3449.

SOKOLOVICZ, B. C. **Microstructure and durability to chlorides of concrete prototypes with rice husk ash with and without prior grinding**. 2013. 164p. Dissertation (Master's Degree in Civil Engineering) - Federal University of Santa Maria, Santa Maria/RS, 2013.

TASHIMA, M. M. **Highly reactive rice husk ash:** production method, physical-chemical characterization and behavior in Portland cement matrices. 2006. Dissertation (Master's Degree in Civil Engineering) - Universidade Estadual Paulista Jùlio de Mesquita Filho, Ilha Solteira, 2006.

UNIVERSOAGRO. **THE energy that comes from farming**. 2013. Available at:

<http://www.uagro.com.br/editorias/agroenergia/2013/11/13/a-energia-que-vem-da lavoura.html>. Acesso em: 12 março 2015.

VOTORANTIM CIMENTOS. **Composition of cements.** Available at:

<http://www.votorantimcimentos.com.br/hotsites/cimento/base.htm>. Accessed on: 08 April15.

YAZIGI, W. **The technique of building**. 12. ed. Sao Paulo: PINI, 2009.

7 APPENDIX A

ABCP dosing method

Values to be determined:

FCJ	25 Mpa
Cement strength at 28 days	32 Mpa
Maximum diameter of coarse aggregate	19 mm
Fineness modulus of fine aggregate	1.9 mm
Compacted unit mass of coarse aggregate	1.69 kg/dm^3
Real specific mass of the fine aggregate	2,600 g/cm3
Specific mass of the coarse aggregate	2.93 g/cm3
SLUMP TEST" Cone Trunk Abatement	80 to 100 mm
Specific mass of cement	3.035 g/cm3

Step 1 - Determining the a/c factor

- Abrans curve for cement

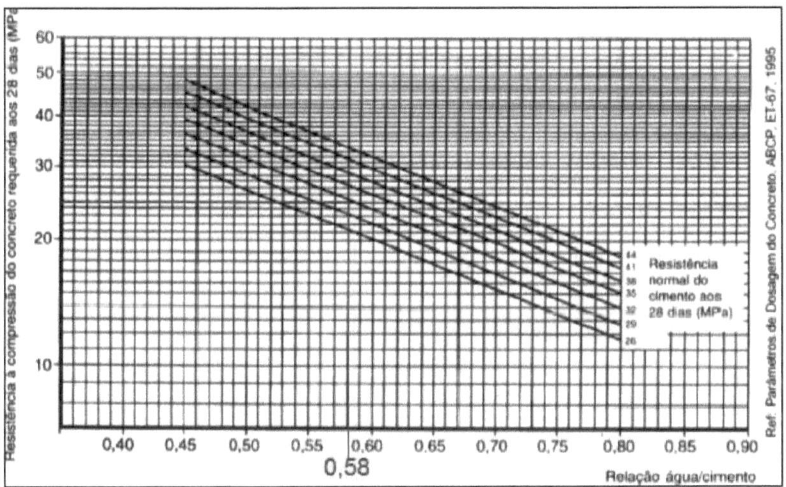

Concrete strength as a function of standard deviation at 28 days

$Fc_{28} = Fck + 1{,}65 \times sd$

$Fc_{28} = 20 + 1{,}65 \times 3$

$Fc_{28} = 24,95$ Mpa

Fator a/c = 0,58

<u>Adopted a/c factor = 0.66</u>

Step 2 - Approximate determination of water consumption (Ca)

Approximate water consumption (l/m $)^3$					
	Maximum weight (mm)				
Sag (mm)	9,5	19	25	32	38
40 a 60	220	195	190	185	180
60 a 80	225	200	195	190	185
80 a 100	230	205	200	195	190

<u>Water consumption = 205 l/m^3</u>

Step 3 - Determining cement consumption (Cc):

$$Cc = \frac{Ca}{a/c} = \frac{205}{0,66}$$
$$Cc = 310,61\, kg/m^3$$

<u>Cement consumption - 310.61 kg/m3</u>

Step 4 - Determining the consumption of coarse aggregate (Cb)

MF	Maximum diameter of coarse aggregate (mm)				
	9,5	19	25	32	38
1,8	0,645	0,77	0,795	0,82	0,845
2	0,625	0,75	0,775	0,8	0,825
2,2	0,605	0,73	0,755	0,78	0,805
2,4	0,585	0,71	0,735	0,76	0,785
2,6	0,565	0,69	0,715	0,74	0,765
2,8	0,545	0,67	0,695	0,72	0,745
3	0,525	0,65	0,675	0,7	0,725
3,2	0,505	0,63	0,655	0,68	0,705
3,4	0,485	0,61	0,635	0,66	0,685
3,6	0,465	0,59	0,615	0,64	0,665

$$Cb = Vb \cdot \gamma_{compactada}$$
$$Cb = 0{,}77 \times 1690$$
$$Cb = 1300 \, kg/m^3$$

<u>Consumption of coarse aggregate - 1300 kg/m^3</u>

Step 5 - Determining the consumption of solid aggregate

- Sand volume (Va)

$$Va = 1 - \left(\frac{cimento}{\delta_{cimento}} + \frac{brita}{\delta_{brita}} + \frac{água}{\delta_{água}} \right)$$
$$Va = 1 - \left(\frac{310{,}61}{3035} + \frac{1300}{2930} + \frac{205}{1000} \right)$$
$$Va = 0{,}255 m^3$$

- Determining the consumption of Miùdo aggregate

$$C_{areia} = V_{areia} \cdot \delta_{areia}$$
$$C_{areia} = 0{,}255 \cdot 2600 \, kg/m^3$$
$$C_{areia} = 663 \, kg/m^3$$

<u>Medium aggregate consumption - 663 kg/m^3</u>

Step 6 - PRESENTING THE DRAWING

Cement : Sand : Gravel 1 : Water/Cement

$$\frac{C_c}{C_c} : \frac{C_{areia}}{C_c} : \frac{C_{brital}}{C_c} : \frac{C_{água}}{C_c}$$
$$1 : \frac{663}{310{,}61} : \frac{1300}{310{,}61} : \frac{205}{310{,}61}$$

$$1 : 2{,}13 : 4{,}19 : 0{,}66$$

Step 7 - Calculating the quantities for the volume = 0.025m^3

Cc = 7,76 kg

Ca = 16,58 kg

Cb = 32,50 kg

H_2O = 5,13 kg

I want morebooks!

Buy your books fast and straightforward online - at one of world's fastest growing online book stores! Environmentally sound due to Print-on-Demand technologies.

Buy your books online at
www.morebooks.shop

Kaufen Sie Ihre Bücher schnell und unkompliziert online – auf einer der am schnellsten wachsenden Buchhandelsplattformen weltweit! Dank Print-On-Demand umwelt- und ressourcenschonend produziert.

Bücher schneller online kaufen
www.morebooks.shop

Printed by Books on Demand GmbH, Norderstedt / Germany